# Local Knowledge, Intellectual Property and Agricultural Innovation

Michael Blakeney • Kadambot H. M. Siddique
Editors

# Local Knowledge, Intellectual Property and Agricultural Innovation

Springer

*Editors*
Michael Blakeney
School of Law
The University of Western Australia
Crawley, WA, Australia

Kadambot H. M. Siddique
Hackett Professor of Agriculture Chair
and Director of The UWA Institute
of Agriculture
University of Western Australia
Crawley, WA, Australia

ISBN 978-981-15-4610-5 ISBN 978-981-15-4611-2 (eBook)
https://doi.org/10.1007/978-981-15-4611-2



This Springer imprint is published by the registered company Springer Nature Singapore Pte Ltd.
The registered company address is: 152 Beach Road, #21-01/04 Gateway East, Singapore 189721, Singapore

# Preface

The Food and Agriculture Organization of the United Nations (FAO) estimates that about 795 million people were chronically undernourished in 2012–14 (FAO, 2015). In 1996, the World Food Summit defined food security as 'when all people at all times have access to sufficient, safe, nutritious food to maintain a healthy and active life' (Maxwell and Buchanan Smith, 1992). With 70% of the world's extremely poor and food insecure people living in rural areas, the role of agriculture – the predominant economic activity in those areas – is crucial for the eradication of poverty and food insecurity.

Smallholder farmers increasingly cultivate marginal lands that are particularly vulnerable to climate change (see Wood et al, 2014). The FAO has observed that with global warming, 'many, of today's poorest developing countries are likely to be negatively affected in the next 50–100 years, with a reduction in the extent and potential productivity of cropland' (FAO, 2003). A 1996 FAO study estimated that the largest reduction in cereal production will occur in developing countries, averaging about 10% (FAO, 1996). Putting this in perspective, a projected 2–3% reduction in African cereal production for 2020 was estimated to put 10 million people at risk. Particularly vulnerable to climate change are those low- to medium-income groups in flood-prone areas who may lose stored food or assets, farmers who may have their land damaged or submerged by a rise in sea level and fishers who may lose their catch to shifted water currents or through flooded spawning areas (IPCC, 2019).

Compounding these problems is the estimate that at the current rate of global population increase, the global demand for cereals will increase by 40% from 1995 to 2020 (Serageldin and Pursley, 2000) and that net cereal imports by developing countries will double to meet the gap between production and demand (Pinstrup-Anderson et al., 1999). Currently, the developing world is a net importer of 88 million tons of cereals a year at a cost of US$14.5 billion.

The consistent policy approach to guarantee food security is to promote technological improvements in agriculture. The massive increases in food productivity in the 30 years between 1960 and 1990, described as the Green Revolution, was achieved by developing high-yielding crop varieties, supported by massive increases in fertiliser and insecticide use. By 1990 it had become apparent that the reliance on chemically nurtured, high yielding crop varieties – which had precipitated the Green Revolution – was no longer economically or environmentally acceptable (Conway and Pretty, 1991). Thus, it was argued that to meet the food security needs of the next 30 years and to create wealth in poor communities, there was a need to increase agricultural productivity on the presently available land, while conserving the natural resource base (Conway, 1997).

Governments introduced hybrid varieties, often developed by multinational life-sciences companies, but these were often vulnerable to pest/disease infestation (see Thorburn, 2015). In response, local knowledge and agricultural practices of traditional farming communities were called upon to underpin sustainable agriculture (Pretty, 1995). As this first chapter will reveal, an important implication for food security is the contribution of traditional farmers to conserving and identifying useful biological material that is embodied in biotechnological innovations. Implementing agricultural advances depends on appropriate legal instruments, enabling and recognising change, and local farmers' knowledge and understanding and adapting scientific knowledge (Winarto et al, 2013).

This book examines the contribution of local knowledge and intellectual property to agricultural innovation and food security. This chapter defines the terms at the heart of this study and reviews the literature concerning the contribution of local knowledge to agricultural innovation, particularly in a time of climate change.

Chapter 2 investigates the role of intellectual property rights, particularly patents, plant variety rights and geographical indications in encouraging agricultural innovation.

Chapter 3 looks at the role of local and traditional knowledge in identifying useful biological materials for the development of agriculture. It discusses the concept of biopiracy and surveys the various international instruments seeking to regulate access to those materials and the sharing of benefits from their utilisation.

Chapter 4 looks at DNA patenting and agriculture, the judicial consideration of DNA patents and the growth in patenting stress-tolerant genes. It discusses the patenting of plant varieties and plant breeding methods.

Chapter 5 looks at the origins of the International Convention for the Protection of New Varieties of Plants (UPOV); the requirement that protected varieties be new, distinct, uniform and stable; eligible registrants for plant variety rights protection; the farmer's privilege to save seed; and the compulsory licensing of protected varieties.

The 1991 version of UPOV extended the protection of the new varieties concept to 'essentially derived varieties' (EDVs). Genetic engineering has made it possible to transfer exogenous DNA to plants, together with mutation breeding (new variations via either irradiation or chemical mutagens) and backcrossing (transferring useful traits of a donor parent variety to a recipient variety). Chapter 6 considers the use of DNA markers to identify traits that could be characterised as EDVs and surveys the resolution of varietal disputes.

Chapter 7 examines the international and national legal frameworks that have fostered the current situation with regards to seeds and local agricultural knowledge in Indonesia. It includes case studies to show how such frameworks are impacting on the practices of farmers and how farmers are dealing with this impact and are developing self-help mechanisms. It argues that standardised regulatory approaches ignore the considerable contribution and innovative nature of local farming practices, which still contribute about 70% of the food produced in the world. The chapter discusses the changes that can be expected from the new Draft Law on the Continuous Cultivation of Agriculture and concludes by pointing to the important influence of two key aspects of the political reformation process in Indonesian society after the end of the military backed Suharto government: decentralisation policies which have been beneficial for local councils and NGOs working on rural issues and the creation of a Constitutional Court, which has issued progressive decisions in cases concerning indigenous peoples and farmers.

Chapter 8 describes the cultivation of local rice varieties in the Tabanan Regency of Java in Indonesia, where the cultivation of traditional varieties is an important activity for rural farming households. In addition to the farmer's daily needs, the production of local rice varieties is becoming important for income-generation for families. Also, several ceremonies related to Balinese culture demand the use of local rice varieties. The Department of Agriculture's division on Agricultural Technology Research (BPTP) in 2016 announced a program called 'self-sufficient seed village' in Bali to allow farmers to share seed within their village. Farmer organisations blend traditional agricultural management with modern methods. Governments or companies wishing to collaborate with farmers for seed provision or other activities are obliged to approach the governance structures established under national legislation.

The Protection of Plant Varieties and Farmers' Rights Act, 2001, the Seeds Act, 1966, and the Geographical Indications of Goods (Registration and Protection) Act, 1999, were enacted to foster agricultural innovation in India. Chapter 9 explores the extent to which this legislation has achieved that objective.

Chapter 10 reports on a study of the effectiveness of Indian farmers' rights and geographical indications legislation in promoting the conservation and utilisation of farmers' varieties of rice in Kerala, South West India, an important rice-producing region. A total of 300 rice farmers were surveyed and the government records of the registration of farmers' varieties and geographical indications were examined. The analysis revealed that the farmers surveyed were either ignorant of the legislation or unsure of its effects. Farmers have not been much involved in the registration of farmers' varieties and have not made any claims in relation to the registered varieties. They also tend to confuse the function of geographical indications protection with the protection of farmers' varieties. This is the first study of the effectiveness of the two Indian statutes in promoting agricultural innovation and the implementation of national legislation seeking to implement the farmers' rights provisions of the International Treaty on Plant Genetic Resources for Food and Agriculture.

The retreat by central governments from research, development and extension in agriculture has contributed to the emergence of farmer-based organisations that have a critical new role in the promotion of more sustainable agricultural systems. Chapter 11 provides background detail on the developments that led to the emergence and evolution of grower groups as crucial players in the agricultural paradigm and describes the critical role played by these groups in sustainable growth and development. The chapter looks at self-help groups in India and Australia.

Much of the research for this book was conducted in the context of an Australia Research Council Discovery Grant DP170100747.

Crawley, WA, Australia Michael Blakeney

Crawley, WA, Australia Kadambot H. M. Siddique

# Contents

# About the Contributors

**Christoph Antons** Professor of Law, Faculty of Business and Law, Newcastle Law School, University of Newcastle, Australia; Affiliated Research Fellow, Max Planck Institute for Innovation and Competition, Munich, Germany; Senior Fellow, Centre for Development Research, University of Bonn, Germany; Visiting Professor, Faculty of Law, University of Technology MARA, Shah Alam, Malaysia. Christoph is currently the Project Leader of the Australian Research Council funded projects 'Building an intellectual property system: The Indonesian experience' and 'Food security and the governance of local knowledge in India and Indonesia'. His recent book publications are *The Routledge Handbook of Asian Law* (2017) and *Intellectual Property, Cultural Property and Intangible Cultural Heritage* (Routledge 2018, with William Logan).

**Ida Ayu Astarini** Associate Professor, Biology Department, Faculty of Mathematics and Natural Sciences, Udayana University, Bali, Indonesia; currently Vice Director for student and academic affairs of Postgraduate Program, Udayana University. She was Head of Master's Degree in Biology Program, Udayana University, from 2014 to 2017. Dr. Astarini completed her BSc at the Faculty of Agriculture, Bogor Agricultural University, West Java, Indonesia, then obtained her Master's and PhD degrees from The Faculty of Natural and Agricultural Sciences, The University of Western Australia, under AusAID scholarship. She became a DIKTI Research Fellow in 2011 at The University of Western Australia and a Fulbright Senior Research Fellow at Department of Horticulture, Texas A & M University, College Station, USA, in 2014. She has 20 years' experience in teaching and research, majoring in plant tissue culture, plant breeding and molecular genetics. Dr. Astarini was also actively involved in teaching in the overseas students program at Udayana University on tropical horticulture and ethnobotany. She has published papers in a number of international journals, and has published five books and a book chapter in Indonesia. Dr. Astarini is also a national interviewer for Indonesia-LPDP scholarship programme.

**Beena Anil Biswas** Risk Manager, The Australian Pesticides and Veterinary Medicines Authority. Beena has also served as Adoption Officer with Sugar Research Australia. At an earlier stage of her carrier, she worked as Research Assistant and casual tutor at the University of Western Australia.

**Michael Blakeney** Winthrop Professor, Faculty of Law, University of Western Australia, and Visiting Professor, University of Ferrara. Michael has held chairs of intellectual property law in both Australia and the UK. He has advised the Asian Development Bank, European Patent Office, FAO, WIPO and a number of university and agricultural research institutes on intellectual property management and has directed several European Commission projects concerned with the establishment of an intellectual property enforcement infrastructure in a number of new EU Member States (Bulgaria, Croatia, Latvia, Poland and Romania) and in a number of EU Applicant States in the West Balkans. Michael has also been an intellectual property expert in a number of projects to assist developing countries in Africa and South East Asia to become members of the WTO. At an earlier stage of his career, he was a project officer in the Asia Pacific Development Bureau of WIPO. Michael has written widely on international intellectual property and is a fellow of the Australian Academy of Social Sciences.

**Made Ria Defiani** Associate Professor, Biology Department, Faculty of Mathematics and Natural Sciences, Udayana University, Bali, Indonesia. She completed her BSc at the Faculty of Agriculture, Bogor Agricultural University, West Java Indonesia, and then obtained her Master's degree from School of Horticulture, University of Western Sydney, Hawkesbury, New South Wales, and Doctoral degree from Faculty of Agriculture, Udayana University, majoring in Biodiversity Conservation. During her Doctoral study, she spent 3 months at Walailak University, Thailand, for a sandwich program under DIKTI Scholarship. Made Ria Defiani teaches plant physiology, horticulture and plant tissue culture. Her major research area is physiology of crop and horticultural plants, and she has published a book and a number of papers in scientific journals.

**Seyed Hossein Jamali** Independent researcher based in Melbourne, Australia. A former research staff of Seed & Plant Certification & Registration Institute (SPCRI), Karaj, Iran. He has done research on application of DNA markers in plant variety protection and published scientific articles in peer-reviewed journals. His recent paper 'Insights into deployment of DNA markers in plant variety protection and registration' is a thorough review of 130 articles from literature on this topic. His educational background (PhD in plant breeding) and previous career piqued his curiosity to continue research on 'impacts of intellectual property rights on agriculture and food security'.

**Jayasree Krishnankutty** Professor of Agricultural Extension at Kerala Agricultural University, India. She has long teaching and research experience. Dr. Jayasree has taken up a number of research projects aimed at rural women's empowerment, micro entrepreneurship, smallholder marketing, natural resources management and sustainable agriculture and conservation of traditional knowledge. Market-led extension has been one of her priority areas and she has co-authored a book *Market led extension – dimensions and tools* which is recommended as a reference book for postgraduate students in Agricultural Extension by ICAR in India. She has also authored chapters in a number of books. She has been the coordinator for capacity building programmes for various types of stakeholders. Dr. Jayasree has attended and presented cases in international programmes on market access for smallholders held in the Netherlands, Ghana, Vietnam and South Africa.

**Made Pharmawati** Associate Professor, Biology Department, Faculty of Mathematics and Natural Sciences, Udayana University, Bali, Indonesia. She did her BSc at Bogor Agricultural University, then studied Plant Cellular Physiology for her MSc at Deakin University, Victoria, Australia, and went to The University of Western Australia for her PhD on Plant Biology. Made Pharmawati was an Endeavour Research Fellow at The University of Western Australia in 2009 and also DIKTI Research Fellow in the same university in 2011. All research during her study solidified her interest in investigating plant diversity using DNA markers. She has over 25 years' experience in DNA profiling studies. Currently she is conducting research in genetic responses of plants to abiotic stress and has a project on seagrass diversity and bioprospecting. As a lecturer, Made Pharmawati has been involved in teaching several subjects related to plant genetics and molecular. Since her involvement in academics, she has a number of publications in reputable international scientific journals.

**Adlinanur F. Prihandiani** Research assistant at the Center for Anthropological Studies in the Department of Anthropology, Faculty of Social and Political Sciences, Universitas Indonesia. She obtained her bachelor's degree in anthropology from Universitas Indonesia in 2017. She is currently a research assistant of the Australian Research Council funded project on 'Food security and the governance of local knowledge in India and Indonesia' and a facilitator for the Directorate of Innovation and Business Incubator funded project 'Disseminating Science Field Shops for Indonesian farmers to cope better with climate change'. Her recent publication is *Nurturance and trust in developing agrometeorological learning in a changing climate: The Science Field Shops in Indonesia* (Routledge 2018, written together with the Science Field Shops' team).

**Rajesh J. Raju** Research Assistant, Communication Centre, Kerala Agricultural University (KAU), India. He is a postgraduate student in economics. He has been working in different externally aided research projects at Kerala Agricultural University since 2013. Dr. Raju has worked in various research projects in the field of Environmental Economics and Agricultural Economics during the initial period of his career in KAU and currently working in the field of Agricultural Extension. His research contributions are mainly in the areas of wetlands and their socio-economic dimensions, the field realities of organic farming and water stress and vulnerability of agricultural lands. Dr. Raju has co-authored a number of publications in the field of social sciences.

**Kadambot H. M. Siddique** Hackett Professor of Agriculture Chair and Director of The UWA Institute of Agriculture, The University of Western Australia. Distinguished Professor, Shenyang Agricultural University; Adjunct Professor, Northwest Agriculture & Forestry University, Lanzhou University, Guangzhou University and Huazhong Agricultural University, China; and Fellow of the Australian Academy of Technological Sciences and Engineering, African Academy of Sciences, Australian Agricultural Institute and Indian National Academy of Agricultural Sciences. Professor Siddique was designated by UN FAO as Special Ambassador for the International Year of Pulses 2016. He has over 30 years' experience in agricultural research, teaching and management in both Australia and overseas. Professor Siddique has developed a national and international reputation in agricultural science especially in the fields of crop physiology, production agronomy, farming systems, genetic resources, breeding research in cereal, grain and pasture legumes and oilseed crops. Professor Siddique has conducted research on adaptation of crops to dryland environments and the phenological, morphological, physiological, biochemical and genetic traits that enable crops to cope with various abiotic stresses. Professor Siddique was the highly cited Researcher 2019 in agricultural science (Web of Science).

**Matthew Tonts** Matthew Tonts is Professor of Geography and Executive Dean of the Faculty of Arts, Business, Law and Education at The University of Western Australia. His research focuses on matters related to regional development in rural and remote areas, particularly in Australia, North America and Asia. This includes work on industry adjustment and economic development, employment, and migration. Professor Tonts has held visiting academic positions in the USA, the UK and China, and an advisor to Australian and international governments on regional policy and economic development.

**Sinta Uli** Former research assistant at the Center for Anthropological Studies in the Department of Anthropology, Faculty of Social and Political Sciences, Universitas Indonesia. She was also a research assistant of the Australian Research Council funded project on 'Food security and the governance of local knowledge in India and Indonesia'. Sinta Uli obtained her master's degree in anthropology from Universitas Indonesia in 2015.

**Yunita T. Winarto** Professor of Anthropology in the Department of Anthropology, Faculty of Social and Political Sciences at Universitas Indonesia and a member of the Cultural Commission of the Indonesian Academy of Sciences. She holds a PhD in anthropology from the Australian National University. Her research and community engagement focuses on the dialectics of scientific and local knowledge, natural resource management in agriculture, and response farming to climate change. She is the project leader of 'Disseminating Science Field Shops for Indonesian farmers to cope better with climate change' and the partner investigator of the Australian Research Council funded project on 'Food security and the governance of local knowledge in India and Indonesia'. Her work has been published in several journals and edited volumes. She is the author of *Seeds of Knowledge: The Beginning of Integrated Pest Management in Java* (Yale Southeast Asia Council, 2004) and the editor of the most recent books on *Food Crises and "False Minds": Why is it still Persisting? (Krisis Pangan dan "Sesat Pikir": Mengapa Masih Berlanjut?* 2016), and *Breaking the Scientific Boundaries: The Intellectual Trajectory of Social-Humanities Professors* (*Meretas Batas Ilmu: Perjalanan Intelektual Guru Besar Sosial Humaniora* (together with Rahayu S. Hidayat and M. Budianta, 2019)).

# Chapter 1
# Introduction

**Michael Blakeney and Kadambot H. M. Siddique**

**Abstract** This chapter defines local knowledge in its relationship with traditional knowledge and intellectual property. It reviews the literature on the role of local knowledge in agricultural innovation and its relationship to scientific knowledge and the Green Revolution. The chapter also examines the role of local knowledge in the protection of geographical indications and in the conservation of seed.

**Keywords** Local knowledge · Traditional knowledge · Indigenous knowledge · Agricultural innovation · Geographical indications

## 1.1 "Local Knowledge", "Traditional Knowledge" and "Intellectual Property"

### *1.1.1 Definitional Efforts*

In an evaluation of the contribution of local knowledge to agricultural innovation and food security, a distinction was drawn between the terms "local knowledge" and "indigenous knowledge" or "traditional knowledge". It is also useful to consider the extent to which this knowledges might be classified as "intellectual property".

Confusion between these terms arose during the negotiations within the Intergovernmental Committee on Intellectual Property and Genetic Resources, Traditional Knowledge and Folklore (IGC) of the World Intellectual Property Organization (WIPO), which is looking to settle three treaties seeking to recognise,

The research in this book was supported by the Australian Research Council under the Discovery Projects funding scheme (Project Number DP170100747).

M. Blakeney (✉)
School of Law, The University of Western Australia, Crawley, WA, Australia
e-mail: michael.blakeney@uwa.edu.au

K. H. M. Siddique
Hackett Professor of Agriculture Chair and Director of The UWA Institute of Agriculture, University of Western Australia, Crawley, WA, Australia

 
M. Blakeney, K. H. M. Siddique (eds.), *Local Knowledge, Intellectual Property and Agricultural Innovation*, https://doi.org/10.1007/978-981-15-4611-2_1

regulate and protect traditional knowledge, traditional cultural expressions and genetic resources (Blakeney, 2016). To assist the IGC in its task of formulating draft texts for these treaties, the WIPO Secretariat prepared a *Glossary of Key Terms Related to Intellectual Property and Genetic Resources, Traditional Knowledge and Traditional Cultural Expressions* (WIPO, 2014). In its *Glossary* the WIPO Secretariat confirmed that "there is as yet no accepted definition of traditional knowledge (TK) at the international level". It draws a distinction between TK "as a broad description of subject matter" which:

> …generally includes the intellectual and intangible cultural heritage, practices and knowledge systems of traditional communities, including indigenous and local communities (traditional knowledge in a general sense or *lato sensu*). In other words, traditional knowledge in a general sense embraces the content of knowledge itself as well as traditional cultural expressions, including: distinctive signs and symbols associated with traditional knowledge

and TK in "international debate", where TK "in the narrow sense" refers to:

> …knowledge as such, in particular the knowledge resulting from intellectual activity in a traditional context, and includes know-how, practices, skills, and innovations. Traditional knowledge can be found in a wide variety of contexts, including: agricultural knowledge; scientific knowledge; technical knowledge; ecological knowledge; medicinal knowledge, including related medicines and remedies; and biodiversity-related knowledge, etc. (citing WIPO, 2001).

A number of problems have been identified with the concepts of "indigenous" and "traditional" knowledge. A political obstacle to their protection is that this has become associated with the right of peoples to self-determination. The 1993 report of Erica-Irene Daes, Special Rapporteur of the UN Sub-Commission on Prevention of Discrimination and Protection of Minorities and Chairperson of the Working Group on Indigenous Populations observed that "the protection of cultural and intellectual property is connected fundamentally with the realisation of the territorial rights and self-determination of indigenous peoples" (Daes, 1993, para.4). These principles are explicitly reaffirmed in Article 31 of the United Nations Declaration on the Rights of Indigenous Peoples 2007 which recognises the rights of indigenous peoples to maintain, control, protect and develop their intellectual property over their cultural heritage, traditional knowledge and traditional cultural expressions (UN, 2007).

To Daes, the concept of "indigenous" embraced the notion of a distinct and separate culture and way of life, based on long-held traditions and knowledge that are connected, fundamentally, to a specific territory. Le Gall pointed out that confining the protection of traditional knowledge to indigenous creations, overlooks the contributions of the creations and knowledge of majority populations that might have arrived in countries as slaves or indentured workers (Le Gall, 2014).

It is problematic to conflate "local" or "traditional" knowledge with "indigenous knowledge", as this has a different meaning in developing Asian countries compared with settler colonies (Dove, 2000). Thus a number of Asian governments have expressed reservations about the applicability of the term "indigenous people", which in their views was more appropriately used in connection with the colonial era in Anglo-American settler colonies (Kingsbury, 1999; Murray, 2000; Benjamin, 2002; Antons, 2007; Persoon, 2009).

### 1.1.2 Legal Transplants

In attempting to fit local, traditional and indigenous knowledge into international legal instruments, it is pointed out that in establishing legal classifications, communities are expected to live up to the expectations of outsiders, especially of lawyers and policy-makers with regards to the "authenticity" of their "traditional lifestyles" (Antons, 2008; Forsyth & Walker, 2008).

It has also been suggested that it is problematic to term "traditional" what is essentially an "admixture of local folk knowledge and extra-local scientific knowledge" (Sillitoe, 2006). This has been described as "peasant science" (Frossard, 1998; see also Winarto, 2004). It is also suggested that the term "traditional" does not do justice to composite systems of agriculture (Dove, 2005) combining subsistence-oriented swiddens with market-oriented cash cropping (Cramb, 2007) and the term "indigenous knowledge", which fails to capture the historic movement and exchange of plant genetic resources among different parts of the world underlying local farmer knowledge and practices (Dove, 2000).

The IGC indicated, as early as its third session a penchant "to leave specific determinations of the boundaries of protectable subject matter up to domestic authorities, and for terminology at the international level to be used more to express a common policy direction" (WIPO, 2002, para. 4). Thus it observed that a general definition of the subject matter of intellectual property (IP) protection, especially at the international level, can be distinguished from the more precise tests that are developed and applied case by case at the national or regional level (WIPO, 2002).

A more fundamental problem with attempts to fit traditional, indigenous and local knowledge within the intellectual property paradigm is that the latter reflects the western conception of separating the object of protection from the environment in which it evolved, what Drahos describes as the "tragedy of commodification" (Drahos, 2014). The attempts by the WIPO IGC to fit these categories of knowledge within treaties providing for their exploitation is not only considered to involve a colonialist style of legal transplantation from industrialised to developing countries (Peukert, 2015), but to provoke intra-community disputes over the sharing of benefits (Forsyth, 2015). Thus Bragdon notes that the focus of discussions in the IGC, concerning crop improvement by farmers, focuses on genetic resources per se, rather than the innovative process where all germplasm, traditional or modern, "is treated as a potential input for direct use or further improvement" (Bragdon, 2013).

Another consequence of the legal transplantations involved in fitting traditional knowledge into the international treaty-making process is that the customary law of traditional communities is invariably removed as being inappropriate (Tobin, 2013) or uncertain (Forsyth, 2013).

## 1.2 Local Knowledge, Innovation and the Informal Economy

The negotiations on treaties dealing with traditional knowledge, traditional cultural expressions and genetic resources in the WIPO IGC have occupied 34 sessions of the IGC since 2001 without any prospect of satisfactory conclusion, at least in the near future (see Blakeney, 2016). In part, this long, drawn-out negotiation is attributable to the political difficulties that some WIPO Member States have in recognising the rights or existence of indigenous peoples. In part, the complexity of these negotiations is attributable to the difficulty of fitting traditional knowledge into the intellectual property paradigm, which is largely based upon a model of incentivised innovation focused on R & D to produce explicit and codified scientific and technical knowledge, the so-called Science, Technology and Innovation (STI) model. This contrasts with the experienced-based mode of innovation based on Doing, Using and Interacting (DUI), which focuses on learning from informal interactions resulting in innovations often with tacit elements (see Jensen, Johnson, Lorenz, & Lundvall, 2007). Of course, innovations that are capable of being protected as a category of intellectual property will contain elements of both models, but the agricultural innovations of local communities are almost entirely based on the DUI model and thus do not easily sit within current intellectual property categories.

Even though the informal economy, including agriculture, is said to exceed 90 percent of the economies of developing countries, definitions of the informal economy are elusive (Charmes, 2016). The concept of "informality" has been dated back to studies of African economies in the early 1970s (Hart, 1973). An International Labour Organization (ILO) report of 1972 defined informal economic activities as a way of doing things characterised by (a) ease of entry; (b) reliance on indigenous resources; (c) family ownership of enterprises; (d) small scale of operation; (e) labour-intensive and adapted technology; (f) skills acquired outside the formal school system; and (g) unregulated and competitive markets (ILO, 1972). These characteristics typify agricultural activities in developing and least developed countries.

Another feature of the informal economy is howit conceives intellectual property. In the formal sector the focus is on the use of the intellectual property system to appropriate the products of innovation activities through patents, trademarks, industrial designs and copyright (Hall et al., 2014). In the informal sector, on the other hand, actors give little consideration to appropriating their returns from innovation and rely on semi-formal or informal approaches to appropriation (Hall et al., 2014). In the informal sector a premium is placed on trust, personal relationships, social beliefs, values and norms and an absence of written agreements. In any event, the actors in the informal sector find it difficult to access formal intellectual property rights in part because they are unaware of the intellectual property system and if they were aware, lack the necessary legal skills and support organisations to secure formal intellectual property rights.

Access by producers in the informal economy to agricultural innovations is often facilitated by communications by agricultural extension officers (Klerkz &

Gildemacher, 2012) and field schools arranged by university agronomists (Winarto et al., 2013).

## 1.3 Utility of Local Knowledge

Agriculture in the developing world is dominated by small-scale farmers, farming in marginal environments, using locally developed agricultural methods. These methods have evolved over time and represent the experiences of farmers interacting with their environments to meet their subsistence needs (see Denevan, 1995). For example, management of the toxicity of cassava in disparate marginal environments depends on the local agronomic knowledge in those environments (Ellen & Soselisa, 2012). Local agricultural knowledge is based on a combination of observation, experimentation with local seed varieties and the testing of new cultivation methods to overcome constraints, such as soil infertility and pest infestations. For example, it has been observed that insect herbivores are less abundant on the ancestors of crops than their domesticated successors (Altieri, 1994; Rosenthal & Dirzo, 1997). Traditional crop management practices have been identified as a rich resource for understanding the interactions between biodiversity and ecosystem function to identify the principles needed to develop more sustainable agricultural systems (Dewalt, 1994).

Altieri pointed out that most traditional agroecosystems share several structural and functional similarities (Altieri, 2004)

- High species numbers
- High structural diversity in time and space
- Exploitation of the full range of local microenvironments
- Maintenance of closed cycles of materials and waste through effective recycling practices
- Complex biological interdependencies, resulting in a high degree of natural pest suppression
- Dependence on local resources and human and animal energy, thereby using low levels of input technology and resulting in positive energy efficiency ratios
- Use of local varieties of crops, wild plants, and animals.

These commonalities make it possible to identify how the dynamics of traditional systems lead to general agricultural improvements. For example, it has been observed that small-scale multiple cropping systems more productive than in monocropping systems (Chang, 1977). This may result from crop interactions, where one crop might release nutrients that benefit the other crop (Vandermeer, 1989). Additionally, it has been established, at least since the Irish Potato Famine, that as a general principle, traditional agroecosystems involving a wide variety of cultivars are less vulnerable to catastrophic crop failure because the diversity of crops offers a variety of defences against vulnerability (Thrupp, 1998).

Food production and distribution systems in developed countries are characterised by industrialised methods of food production and processing, global sources and means of supply and corporate modes of financing and governance. In developing countries, on the other hand, increasing attention has been given to local knowledge involved in constructing alternative food networks (AFNs) (see Goodman & Goodman, 2012). These were defined by Feenstra as being "rooted in particular places" with the aim to be "economically viable for farmers and consumers, use ecologically sound production and distribution practices, and enhance social equity and democracy for all members of the community" (Feenstra, 1997). AFNs include localised food chains (Marsden, Banks, & Bristow, 2000; Renting et al., 2003; Ilbery & Maye, 2005), farmers' markets (Holloway & Kneafsey, 2000; Kirwan, 2006; Brown & Miller, 2008) and community supported agriculture (Allen et al., 2003).

It has been suggested that local foods are inherently healthier, safer and more nutritious than larger-scale agricultural production (Nygård & Storstad, 1998; Cembalo et al., 2015; Pascucci, Dentoni, Lombardi, & Cembalo, 2016; Barbera & Dagnes, 2016). Such networks also confer ecological benefits including reduced food miles and carbon emissions (Kneafsey, Cox, Holloway, et al., 2008), although these positive environmental impacts of AFNs have also been questioned Edwards-Jones et al., 2008; Oglethorpe, 2009. The literature of political economists warn against the over-romanticism of localism (DuPuis & Goodman, 2005; DuPuis, Goodman, & Harrison, 2006; Lombardi et al., 2015) but Tregear cautions that AFNs cannot be properly evaluated until the parameters of the institution are agreed (Tregear, 2011).

## 1.4 Local Knowledge and the Green Revolution

As mentioned above, the Green Revolution of the 1960s and 70s, averted the threatened food shortages thorough the use of high-yielding crop varieties, and massive increases in fertiliser and insecticide use. By the 1990s this had become environmentally unacceptable, but was accepted as necessary to increase agricultural productivity on the available land and meet the food security needs for the next 30 years in the face of climate change. A second, but economically, sustainable, Green Revolution has been called for; described by Gordon Conway, the President of the Rockefeller Foundation, as a 'Doubly-Green Revolution' or an "ecologically-sound replay" of the initial Green Revolution (Conway, 1997). Holt-Giménez, went one further, calling for a 'Thrice Green Revolution' of enhanced productivity, elevated incomes and environmental sustainability (Holt-Giménez, 2002).

One of the criticisms of the original Green Revolution was that farmers were encouraged to cultivate high-yielding introduced varieties, often at the expense of traditional varieties with a loss of agrobiodiversity and detrimental effects on crop resilience (Patel, 2012). The proposed Green Revolution 2.0 is to be grounded on stress-resistant seeds, increasing input use efficiency and using farmer knowledge, which was ignored in the original Green Revolution (Scheinman, 2018).

In India, Professor M.S. Swaminathan proposed an 'Evergreen Revolution' based on biotechnological innovation, with a "farming systems approach" harmonising traditional agricultural knowledge with "frontier science" to create beneficial "ecotechnologies" (Swaminathan, 2006. He predicted increased crop yields with environmental sustainability through:

> (i) integrated gene management; (ii) higher factor productivity, with particular reference to water and nutrients; (iii) precision farming and development of the biological software essential for sustainable agriculture; (iv) bioorganic agriculture combining relevant features of organic farming and biotechnology; (v) biomass utilisation for adding economic value to every part of the biomass; and (vi) knowledge connectivity through internet-aided rural knowledge centres (Swaminathan, 2006, 2302).

Professor Swaminathan put this approach into practical operation when he established the M S Swaminathan Research Foundation[1] in 1988 and a Community Agrobiodiversity Centre (CAbC) at Kalpetta, Wayanad, Kerala in 1997 to revitalise of the *in situ,* on-farm, conservation traditions of tribal communities.[2] Wayanad has the highest concentration of tribe communities in Kerala, forming 17.1% of the total population of the district.[3] In addition it is the location of the Kurichiya, Kuruma and Wayanadan Chetty traditional agricultural communities. The CAbC has conducted numerous studies on the agricultural methods and conservation practices of these communities (e.g. Mathew, 2008) with a view to their introduction into contemporary agriculture. The community approach to the conservation of biodiversity and its strategic placement in a "biodiversity hot spot" enhances the capacity of the CAbC to link agricultural NGOs and self-help groups with local farmers to respond to problems, such as the rapid loss of traditional varieties in rice, vegetables and fruits and the loss of biodiversity in on-farm habitats. Major success of the CAbC are the conservation of traditional varieties of crop plants with food value and Rare, Endemic and Threatened species (RET) in wild flowering plants, providing baseline information on wild foods, yams and RET plant species in the Western Ghats region. Local knowledge has been incorporated into the Promotion of Organic and Low External Input Sustainable Agriculture (LEISA) farming methods.

[1] https://www.mssrf.org/content/history-1, accessed 21 October 2019.

[2] https://www.mssrfcabc.res.in/, accessed 21 October 2019.

[3] The dominant tribes are Kurichiya, Kuruma, Paniya, Adiya and Kattunaikka, with other minor communities namely, Koombaranmar, Kadar, Pulayar, Mannan, Kuravar, Malayan and Thachanadan Moopan also living in the district. See https://www.mssrfcabc.res.in/about-the-centre/location/about-wayanad/, accessed 21 October 2019.

## 1.5 Local Knowledge and Scientific Knowledge

In the high-modernist ideology of development planners, local knowledge was destined to be superseded by scientific knowledge (Cramb, 2007; Ellen, 2007; Scott, 1998). It was envisaged that the 'Green Revolution' hybrid varieties and intensified production would replace traditional cultivars and cultivation practice. However, the paradigm shift in development theory from 'top-down' to 'bottom-up' approaches highlighted the need for community-based natural resource management (Brosius, Tsing, & Zerner, 2005; Li, Dodson, Zhou, et al., 2007). The 1992 Convention on Biological Diversity (CBD), in Article 8j, urged convention parties to "respect, preserve and maintain knowledge, innovations and practices of indigenous and local communities embodying traditional lifestyles relevant for the conservation and sustainable use of biological diversity". Similarly, the 2001 International Treaty on Plant Genetic Resources for Food and Agriculture of the UN Food and Agriculture Organization (FAO), emphasised *in situ* conservation and 'farmers' rights' to equitable benefit sharing and participation in decision making in recognition of past and future contributions to plant conservation and development (Brush, 2007). New approaches to natural resource management received a further boost from the decentralisation policies of developing countries (e.g. Wittayapak & Vandergeest, 2010).

The characteristic agricultural system in the mountainous and hilly regions of Central Africa, Southeast Asia and Latin America is swidden agriculture, also known as slash-and-burn or shifting cultivation (Goldjammer, 1988; Meine, Elok, Niken, & Fahmuddin, 2008; Merz et al., 2009; Fox et al., 2009; Ziegler et al., 2010; Van & Van, 2012; Li et al., 2014).

During the post-war decades, the knowledge and associated natural resource management practices of 'traditional' or 'indigenous' peoples were considered to be outmoded, inefficient and environmentally damaging; this was particularly the case with swidden agriculture. However, more recently, it has been acknowledged that traditional and indigenous perspectives might be more environmentally sustainable than first thought. By maintaining a mosaic of cultivated plots, farmers can harness the natural processes of soil regeneration. Drahos observed that adopting the fire management practices that the Aboriginal peoples of Australia employed for thousands of years could also mitigate the ferocity of the annual wildfires which affect Australia (Drahos, 2014).

## 1.6 Local Knowledge in the Legal Discourse

In legal discourse, local agricultural knowledge attracts the attention principally of IP, environmental and international lawyers. Intellectual property lawyers focus on the knowledge aspects, but discuss its protection from a broad base including cultural practices and rights (von Lewinski, 2008; Gibson, 2005; Lixinski, 2013), human rights (Helfer & Austin, 2011), the right to food and food security (Blakeney, 2009; Sherman, 2013) and the right of access to biological resources (Antons, 2010) and concerns about "biopiracy" arising from imbalances between strong IP rights and weak public benefits for traditional farmers and local holders of knowledge about biodiversity (Blakeney, 2004; Robinson, 2010).

Intellectual property analyses tend to focus on the role of the IP system in agricultural innovation and food security, such as the roles of DNA patenting (Blakeney, 2011, 2016) and plant variety protection (Kolady & Lesser, 2009; Blakeney, 2013; Sanderson, 2013) in securing investment in agricultural research. However, the IP category identified as having the greatest relevance to the protection and utilisation of local knowledge in food and food security is geographical indications (GIs).

## 1.7 Geographical Indications, Local Knowledge and Food

The obligation of countries to protect GIs is contained in Article 22.2 of the World Trade Organization (WTO) Agreement on Trade Related Aspects of Intellectual Property Rights (TRIPS). The TRIPS Agreement defines GIs in Article 22.1 as "indications which identify a good as originating in the territory of a Member, or a region or locality in that territory, where a given quality, reputation or other characteristic of the good is essentially attributable to its geographical origin." The local knowledge of people in a particular region, eg agricultural methods, can provide the decisive linkage between the quality or characteristics of a product and its place of origin.

Geographical indications are particularly advantageous for the producers of agricultural products, allowing them to differentiate their products from general commodity products such as rice, coffee and tea, and thereby enhancing market access (see Evans, 2006; Bramley & Bienabe, 2012; Galtier, Belletti, & Marescotti, 2008). At the same time, several researchers have identified the capacity of GIs to capture premium prices due to the higher value that some consumers attach to products differentiated according to their origin. For example, Babcock reported that Bresse poultry in France received quadruple the commodity price for poultry meat (Babcock & Clemens, 2004); a case study by Gerz and Dupont on Comté cheese in France indicated that French farmers receive an average of 14% more for milk destined for Comté and that since 1990 dairy farms in the Comté area are 32% more profitable than similar farms outside the Comté area (Gerz and Dupont (see also Arfini, 1999). Kireeva et al., examined the use of origin marks in the Peoples Republic of China reporting that the price of 'Zhangqiu Scallion' per kilogram increased from 0.2–0.6 yuan before the use of the origin mark to 1.2–5 yuan after (Kireeva, Xiaobing, & Yumin, 2009). 'Jianlian' lotus seed was registered as a GI in 2006, increasing its price from 26–28 yuan per kilogram to 32–34 yuan per kilogram.

Geographical indications can play an important role in signalling the quality of goods (see Hobbs, 2003; Hobbs & Kerr, 2006; Becker, 2008). They are important for signalling credence attributes, particularly as an origin brand will be underpinned by a registration and certification system. Through the use of geographical indications, producers can communicate to consumers the quality and the associated reputation that has been developed for origin products over time (see Winfree & McCluskey, 2005). This communication can be incentivised by the premium prices attracted by a GI to maintain product quality (Moschini, Menapace, & Pick, 2008).

For the perceived benefits of GI labelling to be realised, such as the promotion of environmental sustainability, there needs to be consumer awareness that origin labelling represents qualities linked to natural and human factors. This ties in with consumer demand for traceability in agrifood products (Murdoch, Marsden, & Banks, 2000). Rural product certification schemes have proliferated since the mid-1990s. They include the certification of organic agriculture, fair-trade certification of products from developing countries, and food produced in compliance with sanitary and traceability protocols (Giraud & Amblard, 2003; Mutersbaugh et al., 2005). Consumers have been identified as placing increasing value on the integrity of food, such as the social and environmental standards involved in the production and processing of agrifood products (Giraud & Amblard, 2003; Hobbs et al., 2005). This is particularly the case following recent food safety crises. As it is not unusual for food to be grown, processed and packaged in different places consumer trust in products has eroded, particularly as a consequence of these crises. Studies indicate a willingness of consumers to pay a premium price to producers who offer transparency in relation to the composition and origin of their products. In situations where uncertainty about quality or safety is elevated, such as in a health crisis, origin labelling can become an important means of inferring product quality, e.g. meat labels after the BSE crisis in Europe (Verbeke & Viaene, 1999; Loureiro & Umberger, 2003; Becker, 2008; Lees, 2003) and dairy product labels after the Chinese Melamine crisis (Xu & Wu, 2010).

Concerns about the safety of agrifoods in China has stimulated an interest in mechanisms for assuring traceability in food chains (Zhao, Finlay, & Kneafsey, 2014). In this context GIs "may convey assumed 'local' (traceability) and 'natural' (nutritiousness and safety) characteristics thereby acting as proxies for quality" (Zhao et al., 2014, at 78).

In Europe, where GIs have been developed the longest, there are some empirically based suggestions that both consumer' and producers have expectations of the quality of origin products in the European market (see Teuber, 2011; Stasi et al., 2011). However, studies indicate that, in shaping the quality of the product, European producers have not necessarily addressed the positive environmental effects of their product formulation; more recently, there has been a 'greening' of product specifications reflecting environmental considerations (See Giovannucci et al., 2009). Thus, GIs "provide the opportunity for territorialisation of environmental-friendly production rules, taking into account local specificities" (Belletti & Marescotti, 2002).

The evolution of the specifications of origin productsis the result of long-standing farming practices involving a composite of agricultural, cultural and environmental practices (see Denevan, 1995). Traditional crop management practices have been identified as a rich resource for understanding the interactions between biodiversity and ecosystem function to identify the principles needed to develop more sustainable agricultural systems (Dewalt, 1994).

The codes of practices, which are collectively adopted by producer associations for the purpose of origin labelling often incorporate biodiversity objectives (Larson, 2007). Biénabe et al., refer to the Rooibos industry in South Africa as an example of an industry that has explicitly considered biodiversity concerns in designing its product specifications (Biénabe, Leclercq, & Maizi, 2009). This is because Rooibos production takes place in a biodiverse and environmentally sensitive area.

With greater knowledge of the interdependence between agricultural products and the local environment, producer associations have a greater awareness of threats to the environment in production practices (Riccheri et al., 2007). Consequently, it has been suggested that the "GI registration process can be expected to have a positive impact upon the key components of ecological embeddedness and, in particular, on the way actors involved in the chain address the ecological elements of food production…." (Belletti & Marescotti, 2002 at 95; Marsden et al., 2000).

van de Kop, Sautier, and Gerz (2006) point out that as the registered Comté PDO specifications limit the intensification of farming, farmers use fewer inputs and the environment is better protected, contributing to maintaining the open landscape of both pastures and woodland that are typical of the Jura region. Profitable traditional livestock raising in the Comté area has limited the loss of pastureland to 7% in the GI-approved area, compared to 18% in the non-GI area.

Belletti and Marescotti (2002) in their empirical study of the European olive oil industry, characterised by an extensive use of GIs—identified this industry as a good example of agriculture with many associated positive environmental impacts such as lower rates of soil erosion, improved fire-risk control, water efficiency, lower pollution and higher levels of biodiversity and genetic diversity in olive-tree varieties.

Lamarque & Lambin, 2015, in a study of cheese producers in the French Alps marketing their cheese as 'Tomme de Savoie' and 'Emmental de Savoie', found that farmers used GIs to attract price premiums and generally adopted environmentally sustainable cropping practices. It was conceded that the data from this study might be skewed by the effect of product subsidies under the European Common Agricultural Policy.

Rural sustainability achieved through the preservation of biodiversity, landscapes, and traditional knowledge may be promoted by the protection of GIs (Barham, 2002). For example, Guerra observed that, in the Mexcal region of Mexico, the Agave sugar needed to make Tequila is cultivated and managed from wild or forest Agave species, which encourages the biodiverse Agave species (Guerra, 2004. See also Bowen & Zapata, 2009). GIs can also serve as a tool for encouraging sustainable agricultural practice by legally limiting the scale of production and production methods. Penker notes that origin products impose an increased responsibility of producers to their place of production (Penker, 2006). Lampkin et al. noted that "organic standards provide a mechanism by which farmers pursuing sustainability goals can be compensated by the market for internalising external costs" (Lampkin, Foster, & Padel, 1999).

Bérard and Marchenay described GIs as a means of "enabling people to translate their long-standing, collective, and patrimonial knowledge into livelihood and income" which may also underpin the maintenance of biodiversity (Bérard & Marchenay, 2009). Numerous authors have pointed out that GIs share many of the characteristics of TK as both seek to preserve communal rights, and like TK GIs can be held in perpetuity for as long as a community maintains the practices that guarantee the distinctive quality of a local product.

One of the justifications for the establishment of an early GIs system for the protection of wines produced in France was the role that they played in preserving agriculture and rural employment in areas that were unsuitable for cereals and other crops (Stanziani, 2004). The maintenance and promotion of rural development has been repeatedly advanced as a justification for GIs (Ray, 1998; Banks & Marsden, 2000; Marsden et al., 2000; Ilbery et al., 2001; Pacciani et al., 2001; Belletti & Marescotti, 2011; Blakeney & Mengistie, 2011).

The creation of local jobs through the protection of GIs is a factor influencing rural exodus (O'Connor and Co, 2005); for example, an increase in employment has been observed for the Comté cheese industry. Kop et al. estimated that the production of Comté cheese generates five times more jobs in processing, maturing, marketing, packing, etc. than its generic equivalent, Emmental, and that migration away from the countryside in the Comté area is only half that of the origin-protected area (van de Kop et al., 2011). At a national level, they estimated that while Comté cheeses account for only 10% of Frence's total cheese output, they are responsible for 40% of the job offers for students trained in cheese-making in vocational schools. Similar results have been identified for origin-protected cheeses supporting the milk supply from cattle in northern Italy and sheep in southern Italy (van de Kop et al., 2006).

## 1.8 Local Knowledge and the *In Situ* Conservation of Seed

A significant knowledge contribution has been made by of traditional farmers in the development of new crop types and biodiversity conservation. The economic value of biological diversity conserved by traditional farmers for agriculture is difficult to quantify because conventional breeding focuses on crosses among elite materials from the breeders' own collections and advanced lines developed in public institutions (e.g. Brush, 1994). An increasingly significant economic value of biodiversity is the extent to which it provides a reservoir of species available for domestication, as well as genetic resources for the enhancement of domestic species. The modern biotechnological revolution has enabled the engineering of desirable genetic traits from useful local species. It is estimated that about 6.5% of all genetic research undertaken in agriculture focuses on germplasm derived from wild species and land races (McNeely, 2001).

Local knowledge is particularly important in the development of farming systems adapted to local conditions, and farming practices. This may enable marginal lands to, be used contributing to food security by creating access to food in remote areas and environmental management by, preventing erosion and maintaining soil fertility, and agrobiodiversity. Traditional ecological knowledge is considered the basis of sustainable agriculture (e.g. Johannes, 1989) and by some as underpinning the modern organic farming movement (IFOAM, 2007).

## References


Allen, P., et al. (2003). Shifting places in the agrifood landscape: The tectonics of alternative agrifood initiatives in California. *Journal of Rural Studies, 19*(1), 61–75.

Altieri, M. A. (1994). *Biodiversity and Pest Management in Agroecosystems*. New York: Haworth Press.

Altieri, M. A. (2004). Linking ecologists and traditional farmers in the search for sustainable agriculture. *Frontiers in Ecology and the Environment, 2*(1), 35–42.

Antons, C. (2007). Traditional knowledge, biological resources and intellectual property rights in Asia: The example of the Philippines. *Forum of International Development Studies, 34*(4), 1–18.

Antons, C. (2008). Traditional cultural expressions and their significance for development in a digital environment: Examples from Australia and Southeast Asia. In C. B. Graber & M. Burri-Nenova (Eds.), *Intellectual property and traditional cultural expressions in a digital environment* (pp. 287–301). Cheltenham, England/Northampton, England: Edward Elgar.

Antons, C. (2010). The role of traditional knowledge and access to genetic resources in biodiversity conservation in Southeast Asia. *Biodiversity and Conservation, 19*(4), 1189–1204.

Arfini, F. (1999). The value of typical products: The case of Prosciutto di Parma and Parmigiano Reggiano cheese. In *67th seminar, European association of agricultural economists, LeMans, October 28–30*.

Babcock, B., & Clemens, R. (2004). *Geographical indications and property rights: Protecting value added agricultural products* (MATRIC Briefing paper 04-MBP 7).

Banks, J., & Marsden, T. K. (2000). Integrating Agri-environment policy, farming systems and rural development: Tir Cymen in Wales. *Sociologia Ruralis, 40*(4), 466–480.

Barbera, F., & Dagnes, J. (2016). Building alternatives from the bottom-up: The case of alternative food networks. *Agriculture and Agricultural Science Procedia, 8*, 324–331.

Barham, E. (2002). Towards a theory of value-based labeling. *Agriculture and Human Values, 19*(4), 349–360.

Becker, T. (2008). European food quality policy: The importance of geographical indications, organic certification and food quality insurance schemes in European countries. *The Estey Centre Journal of International Law and Trade Policy, 10*(1), 111–130.

Belletti G., & Marescotti, A. (2002). *Link between origin-labeled products and rural development*, WP Report 3. Development of Origin-Labeled Products: Humanity, Innovations, and Sustainability (DOLPHINS) project, Le Mans.

Belletti, G., & Marescotti, A. (2011). Origin products, geographical indications and rural development. In E. Barham & B. Sylvander (Eds.), *Labels of origin for food. Local development, global recognition* (pp. 75–91). Wallingford, England: CAB International.

Benjamin, G. (2002). On being tribal in the Malay world. In G. Benjamin & C. Chou (Eds.), *Tribal communities in the Malay world: Historical, cultural and social perspectives* (pp. 7–76). Leiden, Netherlands: Institute of Southeast Asian Studies and International Institute for Asian Studies.

Bérard, L., & Marchenay, P. (2009). *Geographical indications, a contribution to maintaining biodiversity?* (Biosphere Reserves, Technical Notes. 3-2008). Paris: UNESCO.

Biénabe, E., Leclercq, M., & Maizi, P. M. (2009). Le rooibos d'Afrique du Sud: comment la biodiversité s'invite dans la construction d'une indication géographique. *Autrepart, 50*, 117–134.

Blakeney, M. (2004). Bioprospecting and biopiracy. In B. Ong (Ed.), *Intellectual property and biological resources* (pp. 393–424). Singapore, Singapore: Marshall Cavendish.
Blakeney, M. (2009). *Intellectual property rights and food security*. Wallingford, England: CAB International.
Blakeney, M. (2011). Patents and plant breeding: Implications for food security. *Amsterdam Law Forum, 3*(3), 73–88.
Blakeney, M. (2013/2012). Patenting of plant varieties and plant breeding methods. Journal of Experimental Botany 63 1069-1074.
Blakeney, M. (2016). The negotiations in WIPO for international conventions on traditional knowledge and traditional cultural expressions. In J. C. Lai & A. M. Dominicé (Eds.), *Intellectual property and access to Im/material goods* (pp. 227–256). Cheltenham, England: Edward Elgar.
Blakeney, M., & Mengistie, G. (2011). Intellectual property and economic development in sub-Saharan Africa. *The Journal of World Intellectual Property, 14*(3–4), 238–264.
Bowen, S. C., & Zapata, A. V. (2009). Geographical indications, terroir, and socioeconomic and ecological sustainability: The case of tequila. *Journal of Rural Studies, 25*(1), 108–119.
Bragdon, S. (2013). *Small scale farmers: The missing element in the WIPO-IGC draft articles on genetic resources, Geneva* (QUNO, Briefing Paper No. 1).
Bramley, C., & Bienabe, E. (2012). Developments and considerations around geographical indications in the developing world. *Queen Mary Journal of Intellectual Property, 2*(1), 14–37.
Brosius, J. P., Tsing, A. L., & Zerner, C. (Eds.). (2005). *Communities and conservation. Histories and politics of community-based natural resource management*. Lanham, MD: Alta Mira Press.
Brown, C., & Miller, S. (2008). The impacts of local markets: A review of research on farmers' markets and community supported agriculture. *American Journal of Agricultural Economics, 90*(5), 1296–1302.
Brush, S. (1994). *Providing farmers' rights through in situ conservation of crop genetic resources*. Berkeley, CA: University of California Press.
Brush, S. B. (2007). Farmers' rights and protection of traditional agricultural knowledge. *World Development, 35*(9), 1499–1514.
Cembalo, L., et al. (2015). Rationally local: Consumer participation in alternative food chains. *Agribusiness, 31*(3), 330–352.
Chang, J. H. (1977). Tropical agriculture: Crop diversity and crop yields. *Economic Geography, 53*(3), 241–254.
Charmes, J. (2016). The informal economy: Definitions, size, contribution and main characteristics. In E. Kraemer-Mbula & S. Wunsch-Vincent (Eds.), *The informal economy in developing nations. Hidden engine of innovation* (pp. 13–42). Cambridge, UK: Cambridge University Press.
Conway, G. (1997). *The doubly green revolution- food for all in the twenty-first century*. Harmondsworth, England: Penguin.
Cramb, R. A. (2007). *Land and longhouse: Agrarian transformation in the uplands of Sarawak*. Copenhagen, Denmark: NIAS Press.
Daes, E-I. (1993). *Discrimination against indigenous peoples*. E/CN.4/Sub.2/1993/28, 28 July.
Denevan, W. M. (1995). Prehistoric agricultural methods as models for sustainability. *Advances in Plant Pathology, 11*, 21–43.
Dewalt, B. R. (1994). Using indigenous knowledge to improve agriculture and natural resource management. *Human Organization, 53*(2), 123–131.
Dove, M. R. (2000). The life-cycle of indigenous knowledge, and the case of natural rubber production. In R. Ellen, P. Parkes, & A. Bicker (Eds.), *Indigenous environmental knowledge and its transformations* (pp. 209–215). Amsteldijk, The Netherlands: Harwood Press.
Dove, M. R. (2005). Use of global legal mechanisms to conserve local biogenetic resources: Problems and prospects. In M. R. Dove, P. E. Sajise, & A. A. Doolittle (Eds.), *Conserving nature in culture: Case studies from Southeast Asia* (pp. 279–305). New Haven, CT: Yale Southeast Asian Studies.
Drahos, P. (2014). *Intellectual property, indigenous people and their knowledge*. Cambridge: Cambridge University Press.

DuPuis, M., & Goodman, D. (2005). Should we go "home" to eat? Toward a reflexive politics of localism. *Journal of Rural Studies, 21*(3), 359–373.

DuPuis, M., Goodman, D., & Harrison, J. (2006). Just values or just value? Remaking the local in agro-food studies. In T. Marsden & J. Murdoch (Eds.), *Between the local and the global* (pp. 241–268). Bingley, UK: Emerald.

Edwards-Jones, G., et al. (2008). Testing the assertion that 'local food is best': The challenges of an evidence-based approach. *Trends in Food Science and Technology, 19*(5), 265–274.

Ellen, R., & Soselisa, H. L. (2012). A comparative study of the socio-ecological concomitants of cassava (Manihot esculenta Crantz) diversity, local knowledge and management in Eastern Indonesia. *Ethnobotany Research and Applications, 10*, 15–35.

Evans, P. (2006). Geographic indications, trade and the functioning of markets. In M. Pugatch (Ed.), *The intellectual property debate: Perspectives from law* (Economics and political economy) (pp. 345–360). London: Edward Elgar.

Feenstra, G. (1997). Local food systems and sustainable communities. *American Journal of Alternative Agriculture, 12*, 28–36.

Forsyth, M. (2013). How can the theory of legal pluralism assist the traditional knowledge debate? *Intersections: Gender and Sexuality in Asia and the Pacific, 33*, 1–41.

Forsyth, M. (2015). Making room for magic in intellectual property policy. In P. Drahos, G. Ghidini, & H. Ullrich (Eds.), *Kritika: Essays on intellectual property* (Vol. 1, pp. 114–133). Cheltenham, UK: Edward Elgar.

Forsyth, T., & Walker, A. (2008). *Forest guardians, forest destroyers: The politics of environmental knowledge in northern Thailand*. Chiang Mai, Thailand: Silkworm Books.

Fox, J., et al. (2009). Policies, political-economy and Swiddens in Southeast Asia. *Human Ecology, 37*(3), 305–322.

Frossard, D. (1998). Peasant science: A new paradigm for sustainable development? *Research in Philosophy and Technology, 17*, 111–126.

Galtier, F., Belletti, G., & Marescotti, A. (2008). *Are geographical indications a way to "decommodify" the coffee market?* Paper presented at by 12th congress of the European Association of Agricultural Economists, August 26–29, Ghent, Belgium.

Gibson, J. (2005). *Community resources: Intellectual property, international trade, and the protection of traditional knowledge*. London: Ashgate Publishing.

Giovannucci, D., et al. (2009). *Guide to geographical indications. Linking products and their origin*. New York: International Trade Center.

Giraud, G., & Amblard, C. (2003). What does traceability mean for beef meat consumer? *Sciences des Aliments, 23*(1), 40–46.

Goldjammer, J. G. (1988). Rural land-use and wildland fires in the tropics. *Agroforestry Systems, 6*(1), 235–253.

Goodman, D., & Goodman, M. K. (2012). *Alternative food networks knowledge, practice, and politics*. Florence, Italy: Taylor and Francis.

Guerra, J. L. (2004). Geographical indications and biodiversity: Bridges joining distant territories. *Bridges, 8*(2), 17–18.

Hall, B., et al. (2014). The choice between formal and informal intellectual property: A review. *Journal of Economic Literature, 52*(2), 375–423.

Hart, K. (1973). Informal income opportunities and urban employment in Ghana. *The Journal of Modern African Studies, 11*(1), 61–89.

Helfer, L. R., & Austin, G. W. (2011). *Human rights and intellectual property: Mapping the global Interface*. Cambridge, UK: Cambridge University Press.

Hobbs, J. E. (2003). Information, incentives and institutions in the agri-food sector. *Canadian Journal of Agricultural Economics, 51*(3), 413–429.

Hobbs, J. E., & Kerr, W. A. (2006). Consumer information, labelling and international trade in Agri-food products. *Food Policy, 31*(1), 78–89.

Hobbs, J. E., et al. (2005). Traceability in the Canadian red meat sector: Do consumers care? *Canadian Journal of Agricultural Economics, 53*(1), 47–65.

Holloway, L., & Kneafsey, M. (2000). Reading the space of the farmers' market: A preliminary investigation from the UK. *Sociologia Ruralis, 40*(3), 285–299.
Holt-Giménez, E. (2002). Measuring farmers' agroecological resistance after hurricane Mitch in Nicaragua: A case study in participatory, sustainable land management impact monitoring. *Agriculture, Ecosystems & Environment, 93*(1–3), 87–105.
IFOAM, (International Federation of Organic Agriculture Movements). (2007). *Principles of organic farming*. Bonn, Germany: IFOAM.
Ilbery, B., & Maye, D. (2005). Alternative (shorter) food supply chains and specialist livestock products in the Scottish-English borders. *Environment and Planning, 37*(5), 823–844.
Ilbery, B., et al. (2001). Quality, imagery and marketing: Producer perspectives on quality products and services in the lagging rural regions of the European Union. *Geografiska Annaler, 83*(1), 27–40.
ILO. (1972). Employment, incomes and equality. In *A strategy for increasing productive employment in Kenya*. Geneva, Switzerland: ILO, Kenya.
Jensen, M. B., Johnson, B., Lorenz, E., & Lundvall, B. A. (2007). Forms of knowledge and modes of innovation. *Research Policy, 36*(5), 680–691.
Johannes, R. E. (1989). *Traditional ecological knowledge: A collection of essays*. Gland, Switzerland: IUCN.
Kingsbury, B. (1999). The applicability of the international legal concept of "indigenous peoples" in Asia. In J. R. Bauer & D. A. Bell (Eds.), *The east Asian challenge for human rights* (pp. 336–377). Cambridge: Cambridge University Press.
Kireeva, I., Xiaobing, W., & Yumin, Z. (2009). *Comprehensive feasibility study for possible negotiations on a geographical indications agreement between China and the EU*. Brussels: EU-China IP2.
Kirwan, J. (2006). The interpersonal world of direct marketing: Examining conventions of quality at UK farmers' markets. *Journal of Rural Studies, 22*(3), 301–312.
Klerkz, L., & Gildemacher, P. (2012). *The role of innovation brokers in agricultural innovation systems, in World Bank, agricultural innovation systems: An investment sourcebook* (pp. 211–230). Washington, DC: World Bank.
Kneafsey, M., Cox, R., Holloway, L., et al. (2008). *Reconnecting consumers, producers and food: Exploring alternatives*. Berg: Oxford.
Kolady, D. E., & Lesser, W. (2009). Does plant variety protection contribute to crop productivity? Lessons for developing Countries from US wheat breeding. *The Journal of World Intellectual Property, 12*(2), 137–152.
Lamarque, P., & Lambin, E. F. (2015). The effectiveness of marked-based instruments to foster the conservation of extensive land use: The case of geographical indications in the French Alps. *Land Use Policy, 42*, 706–717.
Lampkin, N., Foster, C., & Padel, S. M. (1999). *The policy and regulatory environment for organic farming in Europe: Country reports, organic farming in Europe: Economics and policy* (Vol. 2). Hohenheim: Die Deutsche Bibliothek – CIP-Einheitsaufnahme.
Larson, J. (2007). *The relevance of geographical indications and designations of origin for the sustainable use of genetic resources*. Global Facilitation Unit for Underutilised Species: Rome.
Le Gall, S. B. (2014). Intellectual property, traditional knowledge and cultural property protection. In *Cultural signifiers in the Caribbean and the Americas*. London: Routledge.
Lees, M. (2003). *Food authenticity and traceability*. Cambridge: Woodhead Publishing.
Li, X., Dodson, J., Zhou, X., et al. (2007). Early cultivated wheat and broadening of agriculture in Neolithic China. *The Holocene, 17*(5), 555–560.
Li, P., et al. (2014). A review of Swidden agriculture in Southeast Asia. *Remote Sensing, 6*, 1654–1683.
Lixinski, L. (2013). *Intangible cultural heritage in international law*. Oxford: Oxford University Press.
Lombardi, A., et al. (2015). Are good guys more likely to participate in local agriculture? *Food Quality and Preference, 45*, 158–165.

Loureiro, M. L., & Umberger, W. J. (2003). Estimating consumer willingness to pay for country-of-origin labelling. *Journal of Agricultural and Resource Economics, 28*(2), 287–301.
Marsden, T., Banks, J., & Bristow, G. (2000). Food supply chain approaches: Exploring their role in rural development. *Sociologia Ruralis, 40*, 424–438.
Mathew, E. (2008). *A preliminary study on the biodiversity conservation practices of Wayanadan Chetty of Wayanad District, Kerala*. Kalpetta: M.S. Swaminathan Research Foundation, Community Agrobiodiversity Centre.
McNeely, J. A. (2001). Biodiversity and agricultural development: The crucial institutional issues. In D. R. Lee & C. B. Barrett (Eds.), *Tradeoffs or synergies? Agricultural intensification, economic development and the environment* (pp. 399–419). Wallingford: CABI.
Meine, V. N., Elok, M., Niken, S., & Fahmuddin, A. (2008). *Swiddens in transition: Shifted perceptions on shifting cultivators in Indonesia*. World Agroforestry Centre: Bogor.
Merz, O., et al. (2009). Swidden change in Southeast Asia: Understanding causes and consequences. *Human Ecology, 37*(3), 259–264.
Moschini, G., Menapace, L., & Pick, D. (2008). Geographical indications and the provision of quality in agricultural markets. *American Journal of Agricultural Economics, 90*(3), 794–812.
Murdoch, J., Marsden, T., & Banks, J. (2000). Quality, nature, and embeddedness: Some theoretical considerations in the context of the food sector. *Economic Geography, 76*(2), 107.
Murray, L. T. (2000). Locating indigenous environmental knowledge in Indonesia. In R. Ellen, P. Parkes, & A. Bicker (Eds.), *Indigenous environmental knowledge and its transformations: Critical anthropological perspectives* (pp. 121–149). London: Routledge.
Mutersbaugh, T., et al. (2005). Editorial. Certifying rural spaces: Quality certified products and rural governance. *Journal of Rural Studies, 21*(4), 381–388.
Nygård, B., & Storstad, O. (1998). De-globalization of food markets? Consumer perceptions of safe food: The case of Norway. *Sociologia Ruralis, 38*(1), 35–53.
O'Connor and Co. (2005). *Geographical indications and the challenges for ACP countries*. Agritrade: Brussels.
Oglethorpe, D. (2009). Food miles – the economic, environmental and social significance of the focus on local food. *CAB Reviews: Perspectives in Agriculture, Veterinary Science, Nutrition and Natural Resources, 4*(072), 1–11.
Pacciani, A. et al. (2001). *The role of typical products in fostering rural development and the effects of regulation (EEC)*. 2081/92, 73rd Seminar of the European Association of Agricultural Economists Ancona 28–30 June.
Pascucci, S., Dentoni, D., Lombardi, A., & Cembalo, L. (2016). Sharing values or sharing costs? Understanding consumer participation in alternative food networks. *NJAS – Wageningen Journal of Life Sciences, 78*, 47–60.
Patel, R. (2012). The long green revolution. *The Journal of Peasant Studies, 40*(1), 1–63.
Penker, M. (2006). Mapping and measuring the ecological embeddedness of food supply chains. *Geoforum, 37*, 368–379.
Persoon, G. A. (2009). Being indigenous' in Indonesia and the Philippines. In C. Antons (Ed.), *Traditional knowledge, traditional cultural expressions and intellectual property law in the Asia-Pacific region* (pp. 195–216). Alphen aan den Rijn: Kluwer Law International.
Peukert, A. (2015). Intellectual property: The global spread of a legal concept. In P. Drahos, G. Ghidini, & H. Ullrich (Eds.), *Kritika: Essays on intellectual property* (Vol. 1, pp. 84–113). Cheltenham, UK: Edward Elgar.
Ray, C. (1998). Culture, intellectual property and territorial rural development. *Sociologia Ruralis, 38*(1), 3–20.
Renting, H., Marsden, T. K., & J Banks. (2003). Understanding alternative food networks: Exploring the role of short food supply chains in rural development. *Environment and Planning A, 35*(3), 393–411.
Riccheri, M., et al. (2007). *Assessing the applicability of geographical indications as a means to improve environmental quality in affected ecosystems and the competitiveness of agricultural products, IPDEV research report*. Brussels: European Commission.

Robinson, D. F. (2010). *Confronting biopiracy: Challenges, cases and international debates*. London/Washington, DC: Earthscan.

Rosenthal, J. P., & Dirzo, R. (1997). Effects of life history, domestication and agronomic selection on plant defense against insects: Evidence from maizes and wild relatives. *Evolutionary Ecology, 11*(3), 337–355.

Sanderson, J. (2013). Can intellectual property help feed the world? Intellectual property, the PLUMPYFIELD® network and a sociological imagination. In C. Lawson & J. Sanderson (Eds.), *The intellectual property and food project: From rewarding innovation and creation to feeding the world* (pp. 145–173). Farnham, Surrey/Burlington, VT: Ashgate.

Scheinman, E. (2018). *Contested futures – Traditional rice as a keystone of alternative agriculture movements. Agricultural knowledge, seed systems and agrarian reform in South Karnataka, India*. MES thesis, Department of Environmental Studies, Queen's University, Kingston, Ontario, Canada.

Scott, J. C. (1998). *Seeing like a state: How certain schemes to improve the human condition have failed*. New Haven, CT: Yale University Press.

Sherman, B. (2013). Reconceptualizing intellectual property to promote food security. In C. Lawson & J. Sanderson (Eds.), *The intellectual property and food project: From rewarding innovation and creation to feeding the world* (pp. 23–38). Farnham, Surrey/Burlington, VT: Ashgate.

Sillitoe, P. (2006). Ethnobiology and applied anthropology: Rapprochement of the academic with the practical. In R. Ellen (Ed.), *Ethnobiology and the science of humankind* (pp. 147–175). Oxford: Blackwell Publishing.

Stanziani, A. (2004). Wine reputation and quality controls: The origin of the AOCs in 19th century France. *European Journal of Law and Economics, 18*(2), 149–167.

Stasi, A., et al. (2011). Italian wine demand and differentiation effect of geographical indications. *International Journal of Wine Business Research, 23*(1), 49–61.

Swaminathan, M. S. (2006). An evergreen revolution. *Crop Science, 46*(5), 2293–2303.

Teuber, R. (2011). Consumers' and producers' expectations towards geographical indications; empirical evidence for a German case study. *British Food Journal, 113*(7), 900–918.

Thrupp, L. (1998). *Cultivating diversity: Agrobiodiversity for food security*. Washington, DC: World Resources Institute.

Tobin, B. M. (2013). Bridging the Nagoya compliance gap: The fundamental role of customary law in protection of indigenous peoples, law. *Environment and Development Journal, 9*(2), 142–151.

Tregear, A. (2011). Progressing knowledge in alternative and local food networks: Critical reflections and a research agenda. *Journal of Rural Studies, 27*(4), 419–430.

UN. (2007). *United Nations declaration on the rights of indigenous peoples*. GA Res 61/295, UN GAOR, 61st sess, 107th plen mtg, Supp No 49, UN Doc A/RES/61/295 (13 September 2007).

van de Kop, P., Sautier, D., & Gerz, A. (2006). *Origin-based products. Lessons for pro-poor market development*. Amsterdam, The Netherlands: KIT.

Van, V., & Van, N. (2012). Trends, drivers and impacts of changes in swidden agriculture in tropical forest-agriculture frontiers. A global assessment. *Global Environmental Change, 22*(2), 418–429.

Vandermeer, J. (1989). *The ecology of intercropping*. Cambridge, UK: Cambridge University Press.

Verbeke, W., & Viaene, J. (1999). Consumer attitude to beef quality labeling and associations with beef quality labels. *Journal of International Food and Agribusiness Marketing, 10*(3), 45–65.

von Lewinski, S. (2008). *Indigenous heritage and intellectual property: Genetic resources, traditional knowledge and folklore* (2nd ed.). Wolters Kluwer: Alphen aan den Rijn.

Winarto, Y. T. (2004). *Seeds of knowledge: The beginnings of integrated Pest Management in Java*. New Haven, CT: Yale University Southeast Asian Studies.

Winarto, T., et al. (2013). Agrometeorological learning increasing farmers' knowledge in coping with climate change and unusual risks. *Southeast Asian Studies, 2*(2), 323–349.

Winfree, J. A., & McCluskey, J. (2005). Collective reputation and quality. *American Journal of Agricultural Economics, 87*(1), 206–213.

WIPO. (2001). *Intellectual property needs and expectations of traditional knowledge holders WIPO, WIPO report on fact-finding missions on intellectual property and traditional knowledge (1998–1999)*. Geneva, Switzerland: WIPO.

WIPO. (2002). *Secretariat, 'Traditional knowledge – Operational terms and definitions'*. 3rd Session (WIPO Doc. WIPO/GRTKF/IC/3/9).

WIPO. (2014). *Glossary of key terms related to intellectual property and genetic resources, traditional knowledge and traditional cultural expressions*. 21st Session (WIPO Doc. WIPO/GRTKF/IC/28/INF/7).

Wittayapak, C., & Vandergeest, P. (2010). *The politics of decentralization: Natural resource management in Asia*. Chiang Mai, Thailand: Mekong Press.

Xu, L., & Wu, L. (2010). Food safety and consumer willingness to pay for certified traceable food in China. *Journal of the Science of Food and Agriculture, 90*(8), 1368–1373.

Zhao, X., Finlay, D., & Kneafsey, M. (2014). The effectiveness of contemporary geographical indications (GIs) schemes in enhancing the quality of Chinese agrifoods – Experiences from the field. *Journal of Rural Studies, 36*, 77–86.

Ziegler, A. D., et al. (2010). Recognizing contemporary roles of Swidden agriculture in transforming landscape of Southeast Asia. *Conservation Biology, 25*(4), 846–848.

# Chapter 2
# Intellectual Property and Agricultural Innovation

**Michael Blakeney**

**Abstract** This chapter examines the legal context in which agricultural innovation occurs. It looks at the relationship between intellectual property rights and agriculture. It considers those intellectual property laws which are most relevant to agricultural innovation: plant variety rights, patents, geographical indications and trade secrets. It concludes with a consideration of the impacts of biodiversity laws upon agriculture.

**Keywords** Intellectual property rights · Patents · Geographical indications · Trade secrets · Biodiversity laws

## 2.1 Intellectual Property Rights

Intellectual property rights (IPRs) are rights conferred by law in relation to certain defined categories of industrial, scientific, and cultural creativity. The principal policy objectives for conferring IPRs are to provide an incentive for investment in innovation through the creation of a defined period of commercial exclusivity during which research and development (R&D) costs can be recovered, as well as to create a legal infrastructure to encourage technology transfer. At the heart of most technology transfer arrangements are IPRs such as patents, plant variety rights, trademarks and geographical indications which are typically packaged together with associated know how agreements, supplies of raw materials and technical and commercial services such as marketing and advertising (Blakeney, 1989).

IPRs are of crucial importance for modern agriculture. They serve to make R&D in agriculture attractive, by generating tradeable assets and as security in those assets for investment.

M. Blakeney (✉)
School of Law, The University of Western Australia, Crawley, WA, Australia
e-mail: michael.blakeney@uwa.edu.au



M. Blakeney, K. H. M. Siddique (eds.), *Local Knowledge, Intellectual Property and Agricultural Innovation*, https://doi.org/10.1007/978-981-15-4611-2_2

The principal IPRs relevant to agricultural innovation are: (i) patents, which protect inventions; (ii) plant variety rights, which protect the breeding of new and distinct plant varieties; and (iii) trademarks and geographical indications, which facilitate the marketing of products by providing protection for the symbols of their manufacturing or geographic origin. Also relevant, but of lesser significance, are (i) industrial designs, which protect the aesthetic appearance of products, such as agricultural machinery; (ii) layout designs of integrated circuits, which are relevant to smart agricultural technologies; (iii) confidential information law, which protects know-how or trade secrets; and (iv) copyright, which protects works of cultural creativity, such as books, articles, scientific papers, and arrangements of data.

The categories of IPRs are not closed and may be supplemented by international agreements. For example, currently being negotiated within the World Intellectual Property Organization (WIPO), the United Nations specialized agency concerned with IPRs, are the texts of three draft instruments dealing with the creation of possible new rights in traditional knowledge, traditional cultural expressions, and genetic resources.

These categories of intellectual property have become recognized as a consequence of international agreements, creating a more or less internationally harmonised legal regime.

This regime dates back to the 1883 Paris Convention on Industrial Property and the 1886 Berne Convention for the Protection of Literary and Artistic Works. Both of these conventions originated as attempts to protect national innovations from imitation. The Paris Convention created an international regime to protect industrial property, including patents, trademarks and industrial designs. The Berne Convention sought to protect the creations of authors and artists. The conventions sought to establish international norms for the protection of industrial property and copyright works, the most important of which was the 'national treatment principle' that aimed to give works originating in one contracting state, the same protection in each of the other contracting states as granted to works of nationals in those states.

A number of IPR treaties and agreements have since been enacted under the Paris and Berne Conventions as a means of updating them to take account of new industrial and technological developments, and as a means of achieving international harmonization of intellectual property rights (See Ilardi & Blakeney, 2004). The effect of implementing international IPRs obligations through domestic legislation is that countries establish intellectual property regimes with national effect. To enable the global exploitation of products and technologies, treaties, under the auspices of WIPO, facilitate the international registration of some forms of intellectual property. This 'global protection system' consists, principally, of the Madrid Agreement Concerning the International Registration of Marks of 1891 (and the 1989 Protocol thereto) which establishes a centrally administered system of obtaining a bundle of trademark registrations in separate jurisdictions; the 1925 Hague Agreement Concerning the International Registration of Industrial Designs; and the 1970 Patent Cooperation Treaty.

Under the Patent Cooperation Treaty for example, a system is established whereby the registration of a patent in the national patent office of one country or the with the WIPO International Bureau in Geneva, can be used as a multiple means of filing in other national or regional patent offices of signatory countries. As part of this process, WIPO facilitates an international search report which is made available to national offices and identifies published patent documents and technical literature ('prior art') which may have an influence on whether the invention is patentable.

In addition to the global protection system treaties, WIPO Agreements also provide for a number of classifications, which aid the process of registration of intellectual property rights. For example, the Nice Classification, established by the 1957 Nice Agreement is a system of classifying goods and services for the purpose of registering trademarks. The Locarno Classification, established by the Locarno Agreement, 1968, is a similar classification used for the registration of industrial designs.

These classification systems create a harmonized registration regime between those countries that have adopted them. Each of the parties to the Nice Agreement, for example, is obliged to apply the Nice Classification in connection with the registration of marks in order to show the classes of goods or services for which the marks are registered. Use of the Nice Classification is also mandatory for the international registration of marks effected by the International Bureau of WIPO, under the Madrid Agreement.

Similarly, the 1971 Strasbourg Agreement created the International Patent Classification (IPC), which is amended from time to time. This categorizes all technologies into sections, subsections, classes and subclasses. The symbol of at least the subclass or subclasses are carried on patent documents issued by the patent office of the country where the application is filed. This facilitates the identification of all patent documents concerning the fields of technology in which an innovator might be interested and is an important step in avoiding the duplication of research.

A significant landmark in the international intellectual property framework was the 1994 Trade-Related Aspects of Intellectual Property Rights (TRIPS) Agreement (See Blakeney, 1996). The perceived growth in the counterfeiting of trademarked goods and in the piracy of copyright works in the 1970s and 1980s led to the development of the Agreement. This Agreement sought to make the enforcement of intellectual property rights more effective by utilising the sanctions and dispute settlement mechanism of the General Agreement on Tariffs and Trade (GATT). This entailed the inclusion of TRIPS as an agreement within the World Trade Organisation (WTO) system. TRIPS imported key provisions from the Paris and Berne Conventions, as well as including a number of additional obligations. Most importantly, it required a range of sanctions to be made available to rights-holders and for the first time, obliged signatories to require their customs authorities to interdict the trade in industrial quantities of counterfeit trademark goods and pirated copyright works. In 2010 these enforcement provisions were supplemented by the Anti-Counterfeiting Trade Agreement (ACTA) (See Blakeney, 2012a).

Countries which are members of the World Trade Organization (WTO) are obliged to enact and implement intellectual property laws which comply with the provisions of TRIPS. TRIPS requires laws to be enacted on the various categories of intellectual property mentioned above and requires the enforcement of IPRs through a range of civil, administrative and criminal remedies.

A number of other international agreements also contain IPR prescriptions. The most significant of these for agricultural innovation are: (i) the Convention on Biological Diversity 1992 (CBD), which contains obligations of consent of source countries to access to their biological resources and the sharing of benefits when those resources are exploited; and (ii) the International Treaty on Plant Genetic Resources for Food and Agriculture 2001, which contains rules about IPRs arising from the exploitation of the biological resources which are conserved within the Consulting Group for International Agricultural Research (CGIAR).

## 2.2 Intellectual Property Rights and Agriculture

The first legislative proposal for the protection of agricultural innovations was the Papal States Edict of 3 September 1833 concerning the declarations of ownership of new inventions and discoveries in the fields of the technological arts and agriculture (Laclavière, 1962). Agriculture was included in the definition of "industrial property" in the 1883 Paris Convention for the Protection of Industrial Property. Article 1(3) of the Convention declared that

> Industrial property shall be included within the broadest sense and shall apply not only to industry and commerce proper, but likewise to agricultural and extractive industries and to all manufactured or natural products, for example, wines, grain, tobacco leaf, fruit, cattle, minerals, mineral waters, beer, flowers and flour.

Given the state of technology in 1883, the inclusion of these agricultural subjects within the concept of "industrial property" was probably related to the protection of trade marks and indications of source. The first national proposal that foreshadowed the protection of agricultural innovations under patent law was the introduction, in the United States Congress of 1906, of a "Bill to amend the laws of patents in the interest of the originators of horticultural products." This bill was unsuccessful, as were similar bills introduced in 1907, 1908 and 1910. It was not until the Townsend-Parnell Act of 1930, the "Plant Patent Act," that agricultural innovations were recognised by Congress. Legislation similar to the U.S. Plant Patents Act was adopted in Cuba, 1937; South Africa, 1952 and the Republic of Korea, 1973, in an endeavour by those countries to align their patent systems with that of the United States.

As with other categories of IPRs, a key role in the inclusion of agricultural innovations within the international regulatory regime was played by industry associations. The Congrès pomologique de France, held in 1911, called for special protection for plant varieties. The International Union of the Horticultural Profession, also considered the matter at its Congresses in Luxemburg (1911), London (1912) and Ghent (1913). The International Institute of Agriculture in its 1927 Congress had stated that the protection of a denomination was insufficient and that a way had to be found to require "any grower who engaged in reproduction of those breeds for the purposes of sale to pay a royalty to the producer."

The International Federation of Breeders of Staple Crops had, in its 1931 conference, expressed the hope that the legal status of new varieties should be assimilated to that of industrial inventions. Discussions concerning the creation of a new organization to agitate for the promulgation of an international legal regime for the protection of plant varieties occurred at the meetings of the International Breeders' Congress at Leeuwarden in 1936 and the 1937 Conference of the International Organization of Agricultural Industries, also held in the Netherlands. The direct result of these discussions was the foundation in Amsterdam, on November 17, 1938, of the International Association of Plant Breeders for the Protection of Plant Varieties (ASSINSEL). The first ASSINSEL Congress, held in Paris on 8–9 July 1939 adopted a three-point resolution:

- To accept internationally the filing of trademarks and appellations as a means of protection (pending introduction of a patent);
- To adopt the principle of a licence, to be drawn up by ASSINSEL for the purposes of multiplication and sale; and.
- To accept internationally the definition of the word 'original' [as] seed produced, offered or sold by the breeder of the variety or under his control by his licensees or successors in title.

The Second World War interrupted these developments. At its Semmering Congress in June, 1956, a resolution of ASSINSEL called for an international conference to promulgate an international system for the protection of plant varieties. The French Government had been approached by ASSINSEL, because it had indicated a favourable attitude. Invitations were issued to 12 Western European countries to attend a diplomatic conference in Paris, from 7 to 11 May 1957. These developments resulted in the promulgation in 1961 of the International Convention for the Protection of New Varieties of Plants (UPOV), which is discussed below.

In Australia, the Australian Constitution of 1900 in section 51(xviii) had empowered Parliament to legislate on the subjects of "Copyrights, patents of inventions and designs, and trade marks". A threshold consideration of the High Court of Australia was whether agriculture is a proper subject for intellectual property protection. This issue was usefully considered by the Court in *The Grain Pool of WA v The Commonwealth*[1] in which the plaintiff challenged the constitutional validity of the Plant Variety Rights Act 1987 (Cth) and its successor, the Plant Breeders' Rights Act 1994 (Cth). The Grain Pool had received a protected variety of barley, Franklin for the limited purpose of growing trials and malting evaluation. Commercial negotiations with the representative of the Tasmanian proprietor of the variety, had broken down, and it was alleged that the Grain Pool without permission had grown the barley in Western Australia. The Grain Pool's response questioned whether the plant

[1] [2000] HCA 14.

variety rights legislation fell within the definition of intellectual property envisaged by the Australian Constitution.

The Court said that "it would be wrong to regard the legislative grant of monopoly rights in new plant varieties as being, in 1900, outside the 'central type' of the subject of patents of inventions."[2] The High Court noted the comments of Rich J. in the US Court of Appeals decision: *Imazio Nursery, Inc v Dania Greenhouses*[3] who explained that

> "At least as early as 1892, legislation was proposed to grant patent rights for plant-related inventions. Plant patent legislation was supported by such prominent individuals as Thomas Edison who stated that '[n]othing that Congress could do to help farming would be of greater value and permanence than to give to the plant breeder the same status as the mechanical and chemical inventors now have through the law.' It was also supported by Luther Burbank, a leading plant breeder of the day… whose widow stated that her late husband 'said repeatedly that until Government made some such provision [for plant patent protection] the incentive to create work with plants was slight and independent research and breeding would be discouraged to the great detriment of horticulture."[4]

In the context of food security, it is worth bearing in mind the US Supreme Court's qualification in *Graham v John Deere Co*[5] that Congress may not "enlarge the patent monopoly without regard to the innovation, advancement or social benefit gained thereby."

## 2.3 Plant Variety Rights

Plant varieties are protected in most countries by specialist (*sui generis*) legislation modelled on the International Convention for the Protection of Plant Varieties (UPOV). The UPOV Convention was concluded in 1961 with a number of limitations, such as the obligation of countries to decide whether to protect UPOV by sui generis legislation or by patents. This limitation effectively excluded the USA, which already had a patent system in place to deal with plants. The Convention was revised in 1978 with the insertion of a privilege for farmers to harvest protected seed for further planting or trading with other farmers. In a final revision in 1991, which made the UPOV convention attractive to the USA, double protection of plant varieties through patenting and sui generis protection was permitted, the farmers' privilege was restricted to the harvesting of seed in reasonable quantities for re-use on a farmer's own property. The 1991 version of UPOV also extended to protect varieties which were "essentially derived" from protected varieties.

The protection under UPOV is afforded to a "breeder" or persons claiming through the breeder who is defined in Article 1 (iv) of the UPOV Convention as the

[2] [2000] HCA 14 at para.26.

[3] 69 F 3d 1560 (1995).

[4] 69 F 3d 1560 at 1562–1563 (1995).

[5] 383 US 1 at 5–6 (1966).

person who bred, or discovered or developed a variety". "Breeding" is generally defined as including the discovery of a plant together with its use in selective propagation so as to achieve a result.

Generally, under plant variety rights legislation the plant breeder is conferred an exclusive right to do or to licence the following acts in relation to propagating material of the variety:

produce or reproduce the material;
condition the material for the purpose of propagation;
offer the material for sale;
sell the material;
import the material;
export the material;
stock the material for the purposes described above.

The duration of plant variety rights under legislation based on the UPOV Convention is 25 years in the case of trees and vines and 20 years for any other variety.

Plant variety protection is established after a registration process. A plant variety is considered to be registrable, if it has a breeder, is distinct, uniform, stable and has not been or has only recently been exploited. A plant variety is considered distinct if it is clearly distinguishable from any other variety whose existence is a matter of common knowledge. It is uniform if, subject to the variation which may be expected from the particular features of its propagation, it is uniform in its relevant characteristics on propagation. A plant variety is stable if its relevant characteristics remain unchanged after repeated propagation. A plant variety is taken not to have been exploited if it or propagating material has not been sold to another person by or with the consent of the breeder.

Legislation based on the UPOV Convention generally provides for the grantee of plant variety rights to take all reasonable steps to ensure reasonable public access to the plant variety. This requirement is taken to be satisfied if propagating material of reasonable quality is available to the public at reasonable prices, or as gifts to the public, in sufficient quantities to meet demand. An appropriate person may be licensed to sell or produce propagating material of plants of that variety on reasonable terms and conditions. Generally, an exception to the grant of a compulsory license applies in the case of a plant variety which has no direct use as a consumer product.

The protection of plant varieties is a mandatory obligation for signatories of the TRIPS Agreement under Article 27.3(b). Countries are given the option of protecting plant varieties by patents or sui generis protection or by a combination of both, but since the commencement of the TRIPS Agreement in 1995, most countries have tended to adopt the 1991 Act of the UPOV Convention, by way of compliance. Thus, as of 2 May 2019, the UPOV Convention has 75 signatories, with 41 of those joining after 1 January, 1995. Despite numerous commentaries and proposals for the adoption of alternative *sui generis* models, (eg Dhar, 2002; Helfer, 2002; Robinson, 2008) only a few countries have adopted alternatives to UPOV. This is, in particular, the case of India, Malaysia and Thailand that have combined PVP with benefit sharing provisions inspired by the CBD (see Ramanna, 2003; Azmi, 2004; Kanniah, 2005; Ranjan, 2009; Masarek, 2010; Brahmi & Chaudhary, 2011).

One of the reasons why countries have tended to adopt UPOV 1991, rather than to craft a *sui generis* alternative, is that the IPR chapters in the free trade agreements signed since the 1990's by the USA and the EU with their bilateral partners includes the obligation to join UPOV 1991 (Winter, 2010; Kennedy, 2017).

## 2.4 Patents

### *2.4.1 Patent Principles*

A patent is a statutory privilege granted by a government to an inventor and to other persons deriving their rights from the inventor, for a fixed period of years, to exclude other persons from manufacturing, using or selling a patented product or from using a patented method or process. Patent rights are conferred by statute as a matter of right to the person who is entitled to apply for it and who fulfils the prescribed registration requirements. The protection secured by the registration of a patent is usually limited in time. For example, under the UK Patents Act 1977, s.25, the term of protection is 20 years. At the end of the period of protection, the patented invention is said to be within the public domain, that is available for anyone to exploit.

An invention is usually defined as an idea which permits the solution of a specific problem in a field of technology. The applicant for the protection of an invention is usually the inventor or his successor in title. For an invention to be protected by a patent under most systems of laws it must be: (a) new; (b) involve an inventive step; (c) be industrially applicable; and (d) not be a category of excluded invention.

As intellectual property laws take their existence from legislation interpreted through judicial determinations, it is possible for legislation and case determinations to decree that something, which might have been considered a discovery to be an invention. For example, the European Parliament in its Biotechnology Directive provides that that biological material which is isolated from its natural environment or produced by means of a technical process is deemed to be an invention even if this material previously occurred in nature. The US Supreme Court took a similar position in *Diamond* v *Chakrabarty*[6] in ruling that a bacterium genetically engineered to degrade crude oil was an invention.

The modern biotechnological revolution has enabled the engineering of desirable genetic traits from useful local species. Genetic engineering has permitted the expeditious introduction of a wide range of desirable traits into plants. These include:

[6] 447 US 303 (1980).

pest control traits such as insect, virus and nematode resistance as well as herbicide tolerance; post-harvest traits such as delayed ripening of spoilage prone fruits;
agronomic traits such as nitrogen fixation and utilisation, restricted branching, environmental stress tolerance,
male and/or seed sterility for hybrid systems; and.
output traits such as plant colour and vitamin enrichment.

The production of transgenic plants has become possible through the development of a number of enabling and transformation technologies. These technologies, together with the genetic location of beneficial plant traits, have become the subject of patent protection, as a consequence of the favourable decisions of courts in the USA and Europe. A consequence of this for agricultural innovation is that the cultivation, distribution and marketing of crops containing patented genes renders unauthorised persons liable for patent infringement.

## 2.5 Infringement of Agricultural Patents

The unauthorised use of patented material may involve the user in patent liability. In the agricultural sector the potential defendants are; farmers growing seed with patented DNA; traders in proprietary seed; and research institutes utilising patented DNA, or distributing seed with that DNA. Outlined below are a number of the more significant infringement cases.

### *2.5.1 Monsanto Co. v. Scruggs*

The US case *Monsanto Co. v. Scruggs*[7] concerned Monsanto's US patent of a gene which conferred glyphosate resistance for crops such as canola, soybeans and cotton which had the patented gene inserted and which rendered them resistant to Monsanto's glyphosate herbicide "Round Up". Monsanto had licensed this Round Up Ready ("RuR") technology to seed companies and forbade them from selling seed which contained Monsanto's patented genes to growers unless the grower first signed a technology license agreements, reserving the patented technology to Monsanto and which limited purchasers to growing a single commercial crop, i.e. growers could not save seed produced from a harvested crop for replanting during the following growing season.

Mitchell Scruggs, who had not signed a technology licensing agreement, purchased a small quantity of RuR soybeans from a seed company in Memphis. The seed was sufficient to plant approximately ten acres of soybeans. After the fall harvest, Mr. Scruggs retained the soybean seed from those ten acres; he cleaned it and

[7] 342 F. Supp 2d 584 (2004).

saved it for planting during the 1997 crop season. Through saving seed from all subsequent crop seasons up to the year 2000, by 2000, Scruggs had enough saved RuR soybean seed to plant more than 8000 acres. Monsanto claimed that its patent had been infringed by Scruggs.

The Court found in favour of Monsanto, rejecting Scruggs' defence that neither Monsanto's biotechnology nor the plants in their fields were covered by the patent. Scruggs was obviously compromised by the fact that he had directly purchased Monsanto's proprietary technology.

### 2.5.2 *Monsanto Co. v. Schmeiser*

Of greater significance in assessing the impact of patenting on agriculture is illustrated Canadian litigation between Monsanto Canada, Inc. and a farmer, Percy Schmeiser, who had never purchased the patented technology. In 1993, Monsanto had obtained a Canadian patent for glyphosate-resistant plants and it marketed in that country "Roundup Ready Canola." Schmeiser grew canola commercially in Saskatchewan. He had never purchased RuR Canola nor did he obtain a licence to plant it. Yet, in 1998, tests revealed that 95 to 98% of his 1000 acres of canola crop was made up of RuR plants. The origin of the plants is unclear. They may have been derived from RuR seed that blew onto or near Schmeiser's land, and was then collected from plants that survived after Schmeiser sprayed Roundup herbicide around the power poles and in the ditches along the roadway bordering four of his fields.

Monsanto brought an action for patent infringement claiming that by planting glyphosate-resistant seeds Schmeiser was said to use, reproduce and create genes, cells, plants and seeds containing the genes and cells claimed in the plaintiffs' patent.

At the trial of the case Schmeiser argued that by the unconfined release of the gene into the environment Monsanto did not controlled its spread, and did not intend to do so, and they had thus lost or waived their right to exercise an exclusive patent over the gene.[8] Schmeiser further asserted that the patent was invalid and void because:

(a) the alleged invention is a life form intended for human consumption and is not the proper subject matter for a patent; it is self-propagating and can spread without human intervention;
(b) the patent was obtained for an illicit purpose of creating a noxious plant that would spread by natural means to the lands of innocent parties so as to entrap them with nuisance patent infringement claims;
(c) if infringement is found the plaintiffs would in effect obtain a patent for a plant, which it is urged is not possible in Canada in light of the Plant Breeders' Rights

[8] *Monsanto Canada, Inc and Monsanto Company v Percy Schmeiser and Schmeiser Enterprises* 2001 FCT 256, para 12.

Act (PBRA) which provides for protection of new varieties of plants and which preserves the right of a farmer to save and reuse seed.

The trial judge rejected each of these arguments, in finding that Schmeiser had infringed Monsanto's patent. He held that the fact that replication of the gene may occur in the natural course of events, without human intervention after insertion of the gene in the original plant cells, and plants, produced for seed, did not in itself preclude registration as an invention under the Canadian Patent Act the creation of the gene and the process for inserting the gene. The Trial Judge observed that Schmeiser had grown canola from seed which he knew was RuR tolerant. He ruled that the growth of the seed, reproducing the patented gene and cell, and sale of the harvested crop constituted taking the essence of Monsanto's invention, using it, without permission and in so doing infringed the patent.

The case was unsuccessfully appealed to the Federal Court of Appeal which ruled that Schmeiser's saving and planting seed, then harvesting and selling plants that contained the patented cells and genes was an unauthorized use of the patented material.[9]

### 2.5.3 *Monsanto Co. V. McFarlin*

The relationship between patents and plant variety rights was considered by the US Federal Circuit Court in *Monsanto Co. V. McFarling.*[10] Monsanto had required that sellers of its patented seeds should obtain from purchasers a "Technology Agreement," in which they agreed that the seeds were to be used "for planting a commercial crop only in a single season" and that the purchaser would not "save any crop produced from this seed for replanting, or supply saved seeds to anyone for replanting." Mr. McFarling, a farmer in Mississippi, purchased Roundup Ready soybean seed in 1997 and again in 1998; he signed the Technology Agreement. He saved 1500 bushels of the patented soybeans from his harvest during one season, and instead of selling these soybeans as crop he planted them as seed in the next season. He repeated this activity in the following growing season. This saved seed retained the genetic modifications of the Roundup Ready seed. Mr. McFarling did not dispute that he violated the terms of the Technology Agreement but claimed that the contractual prohibition against using the patented seed to produce new seed for planting, when he produced only enough new seed for his own use the following season, violated the seed saving provision of the US Plant Variety Protection Act (PVPA), which permitted farmers to save seeds of plants registered under the Act. The Court applied *Pioneer Hi-Bred International Inc. v. J.E.M. Ag Supply Inc.*[11]

[9] *Monsanto Canada, Inc. v. Schmeiser.* [2004] 1 S.C.R. 902, 2004 SCC 34.

[10] 302 F.3d 1291 (Fed. Cir. 2002).

[11] 200 F.3d 1374 (Fed. Cir. 2000), *cert. granted,* 148 L. Ed. 2d 954 (2001)

declining to limit the patent law by reference to the PVPA and ruled against Mr. McFarling.

### 2.5.4 *Monsanto Technology LLC v Cefetra BV*

Patent infringement can arise from the importation of patented genetic material, even where a patent might not exist in the exporter's country. This is a potential matter of concern for farmers and exporters of agricultural products. However, a group of related European cases have established that the patented gene must be capable of performing its patented function in the exported material. Monsanto in 2008 had brought infringement actions against importers from Argentina of soy meal derived from of its patented glyphosate resistant soy. In the Dutch litigation: *Monsanto Technology LLC v Cefetra BV and the State of Argentina*[12] Monsanto sought an injunction prohibiting the infringement of the patent in all European countries. The importer denied infringement arguing that as a result of the processing of soy beans to produce the meal, the patented DNA was dead material and could not perform its function of expressing the relevant enzyme. This interpretation was supported by a ruling of the European Court of Justice.[13] A similar position had been taken by the UK High Court.[14]

### 2.5.5 *Organic Seed Growers & Trade Ass'n v. Monsanto Co.*

Monsanto's website (accessed 26 September 2019) states that "since 1997, we have only filed suit against farmers 147 times in the United States" (Monsanto, 2019) It suggests that this is "really a small number... when you consider that we sell seed to more than 350,000 American farmers a year". This small risk of litigation did not dissuade a coalition of 38 farmers, seed sellers, and agricultural organizations led by the Organic Seed Growers and Trade Association, to seek declaratory judgments of non-infringement and invalidity in the District Court for the Southern District of New York,[15] with respect to 23 patents owned by Monsanto Co. and Monsanto Technology, LLC (discussed in Blakeney, 2016).

The plaintiffs described themselves as growers, seed selling businesses, and agricultural organizations which grow, use, or sell conventional seeds, and many of whom have organic certification and who did not want to use or sell transgenic seed

[12] District Court of The Hague 249,983/HAZA 05/2885, March 19, 2008.

[13] Case C-428/08.

[14] *Monsanto Technology LLC v Cargill International S.A* [2007] EWHC 2257 (Pat).

[15] *Organic Seed Growers & Trade Ass'n v. Monsanto Co.*, 851 F. Supp. 2d 544 (S.D.N.Y. 2012) (No. 11-CV-2163)

incorporating Monsanto's technologies. Their principal concern was that, given Monsanto's patent enforcement policy, if their crops became contaminated by transgenic seed they could perversely be accused of patent infringement by the company responsible for the transgenic seed that contaminated them.

The plaintiffs' application for a declaration was refused by the District Court on the ground that there was no justiciable dispute between the parties. It noted that there was no evidence that Monsanto had commenced litigation against inadvertent users of patented seed and there was no evidence that any of the plaintiffs had experienced contamination from Monsanto's seed, or had ever been threatened by Monsanto for patent infringement. Shortly after they initiated the lawsuit, the plaintiffs had asked Monsanto for an express undertaking not to sue. While refusing to enter into such an undertaking Monsanto referred the plaintiffs to its website, which contained the statement that "It has never been, nor will it be Monsanto policy to exercise its patent rights where trace amounts of our patented seeds or traits are present in farmer's fields as a result of inadvertent means." Monsanto's attorneys by letter further expanded on the company's absence of any intent to sue persons in the position of the plaintiffs, declaring that:

> Monsanto is unaware of any circumstances that would give rise to any claim for patent infringement or any lawsuit against your clients. Monsanto therefore does not assert and has no intention of asserting patent-infringement claims against your clients. You represent that "none of your clients intend to possess, use or sell any transgenic seed, including any transgenic seed potentially covered by Monsanto's patents." Taking your representation as true, any fear of suit or other action is unreasonable, and any decision not to grow certain crops unjustified.

These representations were also taken into account by the District Court in ruling that there was no imminent dispute between the parties.

These factors were equally influential in the determination of the Court of Appeals that there was no justiciable controversy between the parties.[16]

The Supreme Court refused the grant of certiorari to allow an appeal to it.[17]

### 2.5.6 *Marsh v Baxter*

It should be noted that the various rulings in *Organic Seed Growers & Trade Ass'n v. Monsanto Co* were applicable only to a dispute in the USA. The District Court had noted in that case the inevitability that conventional crops would be contaminated by trace amounts of windblown pollen or seeds from genetically modified crops or other sources.[18] The Court of Appeals also noted Monsanto's acknowledgement that conventional crops could be exposed to "cross-pollination

[16] *Organic Seed Growers & Trade Ass'n v. Monsanto Co.* (Fed. Cir. 2013)

[17] *Organic Seed Growers & Trade Ass'n v. Monsanto Co.* cert. denied, 134 S. Ct. 901 (2014).

[18] *Organic Seed Growers*, 851 F. Supp. 2d at 548.

from nearby fields where biotech crops are grown" and that they "might inadvertently contain traces of Monsanto biotech genes (because, for example, some transgenic seed or pollen blew onto the grower's land)."[19]

The Court of Appeals referred to a study finding that, despite stringent precautionary measures meant to prevent any commingling of modified and conventional seed crops, a large majority of conventional seed samples had become contaminated by Monsanto's Roundup resistance trait.[20] The District Court found that due to contamination,

> …some unlicensed—and unintended—use of transgenic seeds is inevitable. Like any other seeds, transgenic seeds may contaminate non-transgenic crops through a variety of means, including seed drift or scatter, crosspollination, and commingling via tainted equipment during harvest or postharvest activities, processing, transportation, and storage.[21]

The Court of Appeals observed that genetically modified seeds cannot easily be separated from conventional seeds; thus, a grower who harvests and uses or sells contaminated crops risks incurring infringement liability. The Court of Appeals observed that "both parties seem to concede that at a minimum, using or selling patented seeds without a license is potentially infringing activity."[22] Thus for the purposes of the appeal before it the court assumed "(without deciding) that using or selling windblown seeds would infringe any patents covering those seeds, regardless of whether the alleged infringer intended to benefit from the patented technologies."[23]

The general issue of inadvertent contamination was raised in a dispute between neighbouring farmers in Western Australia. Steve Marsh and his wife had entered into a contract for organic certification with the National Association of Sustainable Agriculture (Australia) Ltd. (NASAA) for their farm, Eagle Rest. Their farm shared a boundary with a neighbouring farm operated by Michael Baxter, which was on the opposite side of a 20 m road. In early 2010, Baxter planted a crop of Monsanto's Round up Ready (RuR) GM canola in the paddocks of his farm which were adjacent to Eagle Rest and he had notified Baxter of his intention to do so.. On 29 September 2010 Marsh had hand delivered to Baxter a document entitled 'Notice of Intention to Take Legal Action', which stated, amongst other things that:

1. *The use of genetically modified organisms* in farming, *including GM canola seed* (GMOs) *has the potential to cause catastrophic commercial losses to non-GM farmers and particularly to non-GM farmers* that have been accredited as being organic (or sustainable) farms (*Organic Farmers*) *if GMOs enter upon and contaminate a non-GM farm or non-GM farm production cycle*;

[19] Organic Seed Growers & Trade Ass'n v. Monsanto Co. (Fed. Cir. 2013) at p.13.

[20] Ibid at p.14.

[21] *Organic Seed Growers*, 851 F. Supp. 2d at 548.

[22] *Organic Seed Growers & Trade Ass'n v. Monsanto Co.* (Fed. Cir. 2013) at p.13.

[23] Ibid.

2. *The principal cause of the commercial losses* to Organic Farmers as a consequence of GMOs contaminating a non-GM farm or non-GM farm production cycle *is as a result of the forfeiture of the price premiums attached to the sale of the produce* grown by Organic Farmers and/or the withdrawal of their accreditation as Organic Farmers. There may also be other costs and expenses incurred as a direct consequence of such contamination by GMOs;

On 25 October 2010, Marsh had published in local newspapers notices to the effect that Eagle Rest was a 'GMO Free Area'. In November 2010, Baxter harvested the GM canola by the process of swathing. This involved cutting the not yet fully matured canola plant close to its base. The swathes were then stood to ripen in the paddock for several weeks before being processed by a header to harvest the ripened seeds from each swathe. This was the first time Baxter had swathed his canola instead of direct harvesting. A total of approximately 245 GM canola swathes were blown by the wind and landed on the Marshes' farm in late November/early December 2010.[24]

After inspection of the GM canola swathes, NASAA decertified approximately 70% of Eagle Rest on the basis of an assessment that the RR Canola swathes and seed pods identified on Eagle Rest posed an "unacceptable risk" of contamination under NASAA Standard 3.2.9 which provided that "Organic certification shall be withdrawn where NASAA considers there is an unacceptable risk of contamination from [genetically modified organisms] or their derivatives."

The Marshes sued Baxter for $AU85,000 for the loss arising out of the loss of NASAA/NCO certification for 70% of Eagle Rest, citing negligence and private nuisance as the two causes of action. These actions were dismissed by the Supreme Court of Western Australia[25] and the Court of Appeal of Western Australia.[26] The trial judge observed that the negligence claim "traverses into legally unchartered territory".[27] He said that the duty alleged was novel and faced a conceptual difficulty given the law's reluctance to expand the categories of cases in which economic loss is recoverable.[28] The judge commented that the duty alleged by the Marshes: "to ensure that the Marshes did not suffer loss" was absolute and set far too high in circumstances involving broad-acre farming which was exposed to uncontrollable seasonal weather.[29] If a duty of care had been more specifically formulated, they might have had a better chance of success. For example, from the point of causation, the Marshes' real grievance was Baxter's choice to harvest by swathing,[30] not his decision to grown GM canola, thus the alleged duty of care could have related to

---

[24] Ibid at [660], [662], [669], [686].

[25] Marsh v Baxter [2014] WASC 187

[26] *Marsh v Baxter* [2015] WASCA 169

[27] [2014] WASC 187 at [307].

[28] [2014] WASC 187 at [328]–[330], [336]–[338].

[29] [2014] WASC 187 at [333]–[334], [335].

[30] [2014] WASC 187 at [341]–[343].

Baxter's choice of harvesting method. However, on the facts, it had not been shown that Baxter had acted negligently, either by growing or by swathing RR canola.

The Court of Appeal of Western Australia, by a majority of 2:1 upheld the decision of the court below and dismissed the appeal.[31] The majority judges, observed that it was not in dispute "that in the particular circumstances of this case, the GM plant material that landed on the appellants' farm posed no risk of any genetic trait transfer to any species of crop or produce on the appellants' land".[32] They noted that although some 245 swathes had entered onto the appellants' land only eight volunteer GM canola plants were ever detected in the subsequent growing season and that these had been identified and pulled out by Mr. Marsh, "presumably before they had set seed".[33]

The joint judgement ruled that the appellants did not established that a duty of care was owed in the particular circumstances of this case[34] and that in any event, reasonable foreseeability of the risk of economic loss was not in itself sufficient to generate a duty of care in the circumstances of the case.[35] Even if such a duty was feasible, the joint judges considered it to be too indeterminate on the facts of this case.[36]

The minority judge was the President of the Court of Appeal, who considered that "a reasonable person in the position of the respondent ought to have known that there was a real risk that GM canola swaths could be blown by strong winds" from his property onto Eagle Rest[37] and that "the respondent had no compelling reason to harvest his GM canola in late 2010 by swathing."[38]

No further appeal was taken in Marsh v Baxter to the High Court of Australia. It of course should be noted that the action in this case did not concern IPRs and Monsanto was not the defendant, although it had an obvious commercial interest in the outcome of the litigation. However, Monsanto was the defendant in a Canadian action brought by organic grain farmers concerned about the loss of lucrative food markets in Japan and Europe, which they attributed to the development and commercial introduction into Canada of GM canola.[39] The Saskatchewan court, taking a similar approach to the West Australian courts in *Marsh v Baxter*, ruled that any losses were attributable to standards imposed by organic certifiers and by foreign markets. It also noted a determination by the Canadian Food Inspection Agency that GM canola had not been shown to be harmful to humans.[40]

[31] *Marsh v Baxter* [2015] WASCA 169
[32] Ibid at para 385.
[33] Ibid at para 426.
[34] Ibid at para 745.
[35] Ibid at para 704.
[36] Ibid at para 744.
[37] Ibid at para 135.
[38] Ibid at para 136.
[39] *Larry Hoffman and Others v Monsanto Canada Inc and Bayer Cropscience Inc*. 2005 SQKB 225.
[40] ibid. at para 22.

## 2.6 Patenting of Plant Breeding Methods

An issue which has been raised in some recent European litigation is whether patent protection could extend to protect plant breeding methods. This was tested by the Enlarged Board of Appeal (EBA) of the European Patent Office (EPO) in two determinations (see Blakeney, 2012b). The European Patent Convention (EPC) in Article 53(b) specifically excludes the patenting of "plant or animal varieties or essentially biological processes for the production of plants or animals." Rule 23b(5) of the EPC explains that a process for the production of plants and animals is essentially biological if it consists entirely of natural phenomena such as crossing or selection". This language is replicated in the EU Biotechnology Directive which in Article 4.1 excludes from patentability: (a) plant and animal varieties; and (b) essentially biological processes for the production of plants or animals. Article 2.2 states that a process for the production of plants or animals is essentially biological "if it consists entirely of natural phenomena such as crossing or selection."

The EBA was called to rule upon the validity of two patents, one for the crossing and selection of broccoli and the other for tomatoes. The broccoli patent application was for a "method for selective increase of the anticarcinogenic glucosinolates in brassica species". The tomato patent application concerned a "method for breeding tomatoes having reduced water content and product of the method". Both of the patent applications were opposed by interested parties.

The EBA ruled that a non-microbiological process for the production of plants which contains or consists of the steps of sexually crossing the whole genomes of plants and of subsequently selecting plants is in principle excluded from patentability as being "essentially biological" within the meaning of Article 53(b) EPC (further discussed in chap. 5).

## 2.7 Trade Marks

Trade marks paly a valuable role in the marketing of agricultural products. Trade marks are generally protected by registration. To be registered as a trade mark a sign must be capable of representation in a visible form. Visible signs typically include names, invented or existing words, letters, numbers, pictures and symbols, or combinations of these signs. To be capable of registration a sign must be capable of distinguishing goods or services of one undertaking from those of other undertakings. Excepted from registration in most countries are marks which are not distinctive, or which are deceptively similar to existing marks and marks which violate public order or morality.

The requirement of distinctiveness has been held to disqualify from protection trade marks which are registered designation of plant varieties. For example the attempt to register AR1 as "the name of a registered variety of ryegrass endophyte" was rejected as this was already a registered plant variety and the test applied by the courts was whether a mark is one which other traders are likely in the ordinary course of their business and without any improper motive, to desire to use upon or in connection with their goods.[41]

Registration of a mark confers protection against emulation by traders using identical or substantially similar marks. Most systems of registration permit assignment or licensure. A system of registered user may be provided to record trade mark licences. In the event of infringement of a registered mark, a trade mark proprietor may seek relief in the form of injunction, compensation orders and seizure of infringing goods.

## 2.8 Collective and Certification Marks

A special type of registered trade mark is a collective mark which may be registered by an association whose members may use it if they comply with the requirements fixed in the regulations concerning the use of the collective mark. Thus, the function of the collective mark is to inform the public about certain particular features of the product for which the collective mark is used. An enterprise entitled to use the collective mark may in addition also use its own trade mark. In the USA collective marks are used by agricultural cooperatives of produce sellers. The collective mark owner is an organization which does not sell its own goods, or render services, but promotes the goods and services of its members.

A certification mark may only be used in accordance with the defined standards. The main difference between collective marks and certification marks is that the former may be used only by particular enterprises, for example, members of the association which owns the collective mark, while the latter may be used by anybody who complies with the defined standards.

US State governments typically encourage the registration of certification marks to encourage agricultural producers. For example, the certification mark VIDALIA is owned by the State of Georgia's Department of Agriculture and is "intended to be used by persons authorized by certifier, and … in connection with which it is used are yellow Granex type onions and are grown by authorized growers within the Vidalia onion production area in Georgia as defined in the Georgia Vidalia Onion Act of 1986."[42] Similarly, FLORIDA CITRUS is owned by the State of Florida's Department of Citrus and certifies that the goods bearing the mark "either consist of citrus fruit grown in the State of Florida, under specified standards, or are processed or manufactured wholly from such citrus fruit."[43]

Non-US agricultural producers have also registered certification marks in the USA. For example the Thai Ministry of Commerce of Thailand, has registered

[41] Heritage Seeds Pty Ltd. [2007] Australian Trade Marks Office (ATMO) 4 (25 January 2007).

[42] U.S. Reg. No. 1709019.

[43] U.S. Reg. No. 1559414.

THAI HOM MALI RICE "harvested in Thailand per the standards set by the Ministry of Commerce of Thailand in "Regulations of the Department of Foreign Trade Re: Usage of the Certification Mark of Thai Hom Mali Rice."[44] Similarly, the Tea Board of India has registered DARJEELING to certify "that the tea contains at least 100% tea originating in the Darjeeling region of India and that the blend meets other specifications established by the certifier."[45]

In Europe, the preference is for such marks to be registered as geographical indications.

## 2.9 Geographical Indications

Geographical Indications (GIs) are signs used to designate the place of origin of goods where a given quality, reputation or other characteristic of the good is essentially attributable to its geographical origin. The obligation of countries to protect geographical indications is contained in Article 22.2 of the World Trade Organization (WTO) Agreement on Trade Related Aspects of Intellectual Property Rights (TRIPS).

GIs are particularly advantageous for the producers of agricultural products in allowing them to differentiate their products from general commodity products such as rice, coffee and tea, thereby enhancing market access (Evans, 2006; Bramley & Bienabe, 2012). More than 40 developing countries and LDCs depend on exports of a single agricultural commodity for more than 20% of their total revenues from merchandise exports. For example, Benin depends on cotton for over 80% of its merchandise export earnings. Ethiopia relies on coffee for over 70% of agricultural exports.

A number of researchers have identified the capacity of GIs to capture premium prices because of the higher value that some consumers attach to products differentiated according to their origin. For example, Babcock (2003) reported that Bresse poultry in France received quadruple the commodity price for poultry meat; a case study by Gerz and Dupont (2006) of Comté cheese in France indicated that French farmers receive an average of 14% more for milk destined for Comté and that dairy farms in the Comté area since 1990 are 32% more profitable than similar farms outside the Comté area. Kireeva, Xiaobing, and Yumin (2009), examining the use of origin marks in the Peoples Republic of China, reported that the price of "Zhangqiu Scallion" per kilogram was raised from 0.2–0.6 yuan before the use of the origin mark to 1.2–5 yuan in 2009. "Jianlian" lotus seed was registered as a GI in 2006, leading to a rise in price from 26–28 yuan per kilogram to 32–34 yuan per kilogram.

Among the reasons which have been identified for GI-marked goods attracting premium prices, is that traditionally produced crops are often freer from

---

[44] U.S. Reg. No. 2,816,123.

[45] U.S. Reg. No. 2,685,923.

contaminants, such as herbicides and pesticides and that GIs applied to these goods provides confidence in their safety and traceability (see Blakeney, 2017).

Generally, geographic indications are monitored and protected by producer associations from the relevant region.

Unlike trade marks, geographical indications are not freely transferrable from one owner to another, as a user must have the appropriate connection with the geographical region and must comply with the production practices of that region.

Geographical indications are obtained through registration. A specification is usually filed indicating the relevant geographical area and the product quality characteristics attributable to that area. The application for registration is usually filed by a body representing the producers of that area. This body will also usually be responsible for bringing actions against wrongful users of the geographical indication.

## 2.10 Trade Secrets

Information, such as know-how, which has been originated by a person and which is not in the public domain and in relation to which efforts have been made to keep it confidential may be protected by the law of confidence. It has been held that the theft of genetic material is actionable. For example in *Franklin* v *Giddins,*[46] which was concerned with the theft by a defendant of budwood cuttings from the plaintiffs' orchard which enabled the defendant after grafting to grow early ripening nectarines, was held by the Queensland Supreme Court to involve a theft of confidential information embodied in the genetic composition of the budwood.

In *Pioneer Hi-Bred Int'l v. Holden Found Seeds*[47] the US Eighth Circuit Court of Appeals was concerned with a dispute between competing breeders of corn seed. Pioneer claimed that Holden had developed a seed from misappropriated seed which it claimed were its trade secrets. Holden disputed the genetic similarity between its seed and Pioneer's H3H/H43SZ7. In an attempt to evaluate the parties' competing claims, the court oversaw three series of tests: electrophoresis, reverse phase high-performance liquid chromatography and grow-outs. Each test was supervised by the court, performed by independent experts, and monitored by the parties. The court found that Holden's explanation of the parentage of the seeds was unlikely, preferring Pioneer's theory of parentage, awarding it $US46 million for misappropriation of its trade secrets.

---

[46] (1978) Qd R 72.

[47] 35 F.3d 1226 (eighth Cir. 1994).

## 2.11 Copyright

Copyright law is concerned with the protection and exploitation of the expression of ideas in a tangible form. This can include a table or compilation expressed in words, figures or symbols; and a computer program or compilation of computer programs. Consequently, copyright protection may cover scientific papers, scientific databases, as well as laboratory notebooks, and computer displays of information.

The relevance of copyright law to agricultural research is primarily in the suggestion that copyright might be asserted over the written representation of a gene or amino acid sequence in addition, or as an alternative, to applying for a patent or other intellectual property protection (Kayton, 1982; Derzko, 1993; Coke, 2002; cf. Karnell, 1995).

## 2.12 Biodiversity Laws

As will be discussed in the next chapter, the research activities of agricultural research institutes are increasingly having to take account of the international biodiversity regime. The United Nations Convention on Biological Diversity (CBD), 1992, as supplemented by the Nagoya Protocol, 2010[48] and the International Treaty on Plant Genetic Resources for Food and Agriculture seek to establish an international regime for the conservation and utilisation of the world's biological resources and for the 'fair and equitable sharing' of the benefits arising from the utilisation of those resources.

The traditional knowledge of indigenous peoples throughout the world, has played an important role in identifying biological resources worthy of commercial exploitation (Blakeney, 2002). A significant contribution has also been made by the application of the knowledge of indigenous peoples and traditional farmers in the development of new crop types and biodiversity conservation. These groups have been an important agency in the conservation of plant genetic resources and the transmission of these resources to, plant breeders, research institutes and seed companies. A number of countries, including India, require the sharing of benefits with farmers and traditional communities from the exploitation of those resources.

---

[48] *Nagoya Protocol on Access to Genetic Resources and the Fair and Equitable Sharing of Benefits Arising from their Utilization (ABS) to the Convention on Biological Diversity.*

## References


Azmi, I. M. A. G. (2004). The protection of plant varieties in Malaysia. *The Journal of World Intellectual Property, 7*(6), 877–890.

Babcock, B. A. (2003). Geographical indications, property rights and value-added agriculture. *Iowa Ag Review, 9*(4). article1.

Blakeney, M. (1989). *Legal aspects of technology transfer to developing countries*. Oxford, UK: ESC.

Blakeney, M. (1996). Trade related aspects of intellectual property rights. In *A concise guide to the TRIPs agreement*. London: Sweet & Maxwell.

Blakeney, M. (2002). Intellectual property aspects of traditional agricultural knowledge. In R. E. Evenson, V. Santaniello, & D. Zilberman (Eds.), *Economic and social issues in agricultural biotechnology* (pp. 43–60). Wallingford, UK: CABI Publishing.

Blakeney, M. (2012a). *Intellectual property enforcement. A commentary on the anti-counterfeiting trade agreement (ACTA)*. Cheltenham UK: Edward Elgar.

Blakeney, M. (2012b). Patenting of plant varieties and plant breeding methods. *Journal of Experimental Botany, 63*(3), 1069–1074.

Blakeney, M. (2016). Organic versus GM agriculture in the courtroom in Australia and the USA. In N. Kalaitzandonakes et al. (Eds.), *The coexistence of genetically modified, organic and conventional foods government policies and market practices* (pp. 113–136). New York: Springer.

Blakeney, M. (2017). Geographical indications and environmental protection. *Frontiers of Law in China, 12*(2), 162–173.

Brahmi, P., & Chaudhary, V. (2011). Protection of plant varieties: Systems across countries. *Plant Genetic Resources Journal, 9*(3), 393–403.

Bramley, C., & Bienabe, E. (2012). Developments and considerations around geographical indications in the developing world. *Queen Mary Journal of Intellectual Property, 2*(1), 14–37.

Coke, S. (2002). Copyright and gene technology. *Journal of Law and Medicine, 10*(1), 97–103.

Derzko, N. (1993). Protecting genetic sequences under the Canadian copyright act. *Intellectual Property Journal, 8*, 31–43.

Dhar, B. (2002). *Sui generis systems for plant variety protection. Options under TRIPS*. A discussion paper. Geneva: QUNO.

Evans, P. (2006). Geographic indications, trade and the functioning of markets. In M. Pugatch (Ed.), *The intellectual property debate: Perspectives from law* (Economics and political economy) (pp. 345–360). Cheltenham, UK: Edward Elgar.

Gerz, A., & Dupont, F. (2006). Comté cheese in France: Impact of a geographical indication on rural development. In P. van de Kop, D. Sautier, & A. Gerz (Eds.), *Origin-based products: Lessons for pro-poor market development* (pp. 75–87). Amsterdam, The Netherlands: KIT Publishers.

Helfer, L.R. (2002). *Intellectual property rights in plant varieties: An overview with options for National Governments* (FAO Legal Papers Online #31). Rome: FAO.

Ilardi, A., & Blakeney, M. (2004). *International encyclopaedia of intellectual property treaties*. Oxford: Oxford University Press.

Kanniah, R. (2005). Plant variety protection in Indonesia, Malaysia, the Philippines and Thailand. *The Journal of World Intellectual Property, 8*(3), 283–310.

Karnell, G. (1995). Protection of results of genetic research by copyright or design rights? *EIPR, 17*, 357–362.

Kayton, I. (1982). Copyright in genetically engineered works. *The George Washington International Law Review, 10*(2), 191–218.

Kennedy, M. (2017). Export restrictions in plant breeder's rights. *Journal of International Economics, 20*(4), 883–903.

Kireeva, I., Xiaobing, W., & Yumin, Z. (2009). *Comprehensive feasibility study for possible negotiations on a geographical indications agreement between China and the EU, EU-China IP2*. Brussels, Belgium: EC.

Laclavière, B. (1962). *La protection des droits des obtenteurs sur les nouvelles espèces ou varieties des plantes et la Convention de Paris du 2 Décembre 1961 pour la protection des obtentions végétales, (April, 1962) No. 168.* Bulletin Technique D'information Des Ingénieurs Des Services Agricoles, cited in Heitz, A. (1991). The History of the UPOV Convention and the Rationale for Plant Breeders' Rights, Paper delivered at UPOV Seminar on the Nature of and Rationale for the Protection of Plant Varieties under the UPOV Convention, Buenos Aires, 26–27 Nov.

Masarek, A. (2010). Treetop view of the cathedral: Plant variety protection in South and Southeast Asian least developed countries. *Emory International Law Review, 24*, 433–467.

Monsanto. (2019). *Saved seed and farmer lawsuits*'. Available at http://www.monsanto.com/news-views/pages/saved-seed-farmer-lawsuits.aspx. Accessed 29 Sept 2019.

Ramanna, A. (2003). *India's plant variety and farmers' rights legislation: Potential impact on stakeholder access to genetic resources* (EPTD Discussion Paper No. 96). Washington, DC: IFPRI.

Ranjan, P. (2009). Recent developments in India's plant variety protection, seed regulation and linkages with UPOV's proposed membership. *The Journal of World Intellectual Property, 12*(3), 219–243.

Robinson, D. (2008). Sui generis plant variety protection systems: Liability rules and non-UPOV systems of protection. *Journal of Intellectual Property Law & Practice, 3*(10), 659–665.

UPOV. (1987). *The history of plant variety protection, in, UPOV, the first twenty-five years of the international convention of the protection of new varieties of plants*. Geneva, Switzerland: UPOV.

Winter, L. (2010). Cultivating farmers' rights: reconciling food security, indigenous agriculture, and TRIPs. (Agreement on Trade-Related Aspects of Intellectual Property Rights). *Vanderbilt Journal of Transnational Law, 43*(1), 223–254.

# Chapter 3
# Access to Plant Genetic Resources for Food and Agriculture

**Michael Blakeney**

**Abstract** This chapter looks at the contribution of plant genetic resources to agricultural innovation and concerns about the appropriation of those resources by unauthorized persons. It details 'biopiracy' episodes which have involved patents and plant variety rights. The chapter looks at the role of the traditional knowledge of indigenous peoples and farmers in identifying useful genetic resources. The international conventions regulating access to genetic resources are described, including the Convention on Biological Diversity, the Nagoya Protocol, the International Treaty on Plant Genetic Resources for Food and Agriculture and the World Trade Organization Agreement on Trade Related Aspects of Intellectual Property Rights (TRIPS). The chapter concludes with an examination of the negotiations for a treaty on genetic resources at the World Intellectual Property Organization.

**Keywords** Genetic resources · Biopiracy · International conventions protecting genetic resources · WIPO international treaty on genetic resources

## 3.1 Biodiversity and Plant Genetic Resources

Seventeen countries, including Australia and India, have been identified as "megadiverse" countries with significant proportions of the world's flora and fauna species (Mittermeier et al., 1989). This biodiversity is a valuable repository of genetic material which can be used for agricultural innovations, particularly in a situation of climate change and population growth.

It has been repeatedly observed that crop wild relative species tend to contain greater genetic variation than and thus represent a reservoir of useful variation for

M. Blakeney (✉)
School of Law, The University of Western Australia, Crawley, WA, Australia
e-mail: michael.blakeney@uwa.edu.au



M. Blakeney, K. H. M. Siddique (eds.), *Local Knowledge, Intellectual Property and Agricultural Innovation*, https://doi.org/10.1007/978-981-15-4611-2_3

crop improvement, especially because of their potential to contribute beneficial traits to crops, such as biotic and abiotic resistance, improved yield and climate adaptability (Hajjar & Hodgkin, 2007; Maxted et al., 2012; Prescott-Allen & Prescott Allen, 1988). By way of example, Dwivedi et al. observed that while many dominant genes for climate adaptation and trait enhancement have been lost during cereal crop domestication, they have been retained in the genome of the wild components of the *Triticeae* gene pools. De Pace, et al. noted that in its natural habitat, wild *Triticeae* species such as *Dasypyrum villosum* (*Dv*), whose genome was exposed to millions of years of climatic and environmental changes, "are now expressing increased heading earliness, density stands and plant biomass" (De Pace et al., 2011). They have suggested that deploying whole and dissected *Dv* nuclear genome in the homoeologous wheat genetic background through interspecific hybridization and introgression "could be a lower cost and effective option to help wheat breeders to merge and select the proper adapted gene pools to sustain the needed yearly grain yield increase" (De Pace et al.)

The agricultural value of plant genetic resources is considerable. It is estimated that genetic materials traceable to developing countries account for more than 95% of the output of the world's top twenty food crops (Chen, 2000, 176). It has been estimated that about 6.5% of all genetic research undertaken in agriculture is focussed upon germplasm derived from wild species and landraces (McNeely, 2001). Crop wild relatives, which are the source of potential crop improvements, exist mainly in situ. Maxted and Kell (2009) estimated that only 2–6% of global gene bank collections comprise crop wild relatives and that of the total number of species, only about 6% have been conserved in ex situ collections. There are around 1700 gene banks and germplasm collections around the world with some 7.4 million accessions of plant genetic resources (Dias, 2015, 7). The Consultative Group on International Agricultural Research (CGIAR), comprising eleven Centres hosting international crop and forage collections, holds about 0.7 million accessions of 3446 species from 612 genera. These centres include: Africa Rice Center, International Center for Tropical Agriculture (Centro Internacional de Agricultura Tropical) (CIAT). International Maize and Wheat Improvement Center (Centro Internacional de Mejoramiento de Maiz y Trigo) (CIMMYT), International Potato Centre, (Centro Internacional de la Papa) (CIP), International Center for Agricultural Research in the Dry Areas (ICARDA); International Crop Research Institute for the Semi-Arid Tropics (ICRISAT) and the International Rice Research Institute (IRRI). These CGIAR collections were established from the mid-1960s from deposits by source countries and by the collecting activities of CGIAR centre researchers, who were welcomed into source countries which were comfortable with the mission of the CGIAR to provide improved seed to farmers in developing countries (Blakeney, 1998).

With the development of recombinant DNA technology in the mid-1970s it became possible for persons to identify and commodify, through patenting and plant variety rights protection, the useful germplasm in both *in situ and ex situ* collections. This has circumscribed the availability of these genetic resources for crop improvement. These accessions have been characterised as "biopiracy" and have generated efforts to create an international legal regime to proscribe unauthorised accessions and to regulate the access to plant genetic resources for food and agriculture, as well as encouraging the conservation of those resources. These developments are discussed below.

It should also be noted that "biopiracy" concerns have also been raised outside the agricultural context in relation to the acquisition of biological resources for the development of medicines. (Blakeney, 1997, 2019; Efferth, 2019; Sharma, Maurya, & Brahmacharimayum, 2018; Srivastava, 2011).

## 3.2 Biopiracy

The person credited with coining the term "Biopiracy" is Pat Mooney (Executive Director of ETC Group (Action Group on Erosion, Technology and Concentration), formerly RAFI (Rural Advancement Foundation International), who defined biopiracy as:

> the appropriation of the knowledge and genetic resources of farming and indigenous communities by individuals or institutions who seek exclusive monopoly control (patents or intellectual property) over these resources and knowledge (ETC, 2005).

in the context of the increasing assertion in the 1970s and 1980s of intellectual property rights over plant germplasm (see Blakeney, 2004; Kloppenburg & Kleinman, 1988; Robinson, 2010). Thus Vandana Shiva, the famous Indian environmental activist, explained that "biopiracy" referred "the use of intellectual property systems to legitimize the exclusive ownership and control over biological resources and biological products that have been used over centuries in non-industrialized cultures" (Shiva, 2001). A less pejorative characterization of this practice is "bioprospecting", defined as "the exploration of biodiversity for commercially valuable genetic and biochemical resources" (UNEP, 2000, para. 6). A vigorous scholarship has characterised bioprospecting as a valuable practice which benefits all farmers (Heald, 2003) and has even contested the existence of biopiracy as "rural legend" (Chen, 2005). This controversy had a North-South dimension as the principal sources of useful germplasm are the developing countries of the tropics, whereas the principal exploiters of that germplasm have tended to be the less biodiverse industrialized countries. On the one hand the bio-exploiters have insisted that genetic resources are the common heritage of mankind, whereas source countries and communities have asserted a right to prior informed consent and the sharing of benefits derived from the exploitation of those resources.

It has been observed that all countries are interdependent in their reliance upon germplasm from other countries. Thus, by way of example, it is estimated that Bangladeshi rice contains four varieties from its own landraces and 229 borrowed landraces and USA rice comprises 219 native landraces and 106 borrowed landraces (Fowler & Hodgkin, 2004). However, a number of high profile "biopiracy incidents" have generated demands for the establishment of an effective international legal regime to mediate access to genetic resources. Those in the agriculture domain are mentioned below.

## 3.3 Patenting

The first notorious example of biopiracy concerned patents granted in 1994 by the United States Patent and Trademarks Office (USPTO) and the European Patent Office (EPO) over Neem (*Azadirachta indica*) extracts by the US corporation W.R. Grace & Company and the United States Department of Agriculture. This patent concerned a method for extracting azadirachtin from neem tree seeds to be used as an insecticide.[1] A coalition of environmental NGOs challenged the patent on grounds that the patent lacked novelty and an inventive step because the fungicidal effect of hydrophobic extracts of neem seeds was known and used for centuries in India, both in Ayurvedic medicine to cure dermatological diseases and in traditional Indian agricultural practice to protect crops from being destroyed by fungal infections (Shiva & Holla-Bhar, 1996). These arguments were accepted both by the United States Patent and Trademarks Office (USPTO) and by the European Patent Office (EPO) in revoking the patent. This case generated a substantial campaign in India and other countries against perceived threats to the sovereignty of countries over their biological resources and despite the revocation of the patent, it has come to be regarded as the quintessential example of biopiracy (Eg see Shiva, 2013).

A second example of biopiracy, also involving the biological resources of India concerned a patent granted by the USPTO in September 1997 to RiceTec, an American company based in Texas, for "Basmati rice lines and grains".[2] Basmati rich been cultivated in northern India, as well as in Pakistan for centuries. It is estimated that Basmati rice is India's primary rice export, being cultivated on between 10 and 15% of the total land area under rice cultivation (Shiva, 2000, 85). In April, 2000 the Indian Government challenged a number of the claims in this patent on the basis that the invention lacked novelty (see Subbiah, 2004, 552–53). The USPTO ruled that most of the patent claims were invalid, but it upheld the patent in relation to three hybrid lines which RiceTec had developed from Basmati.[3] A separate complaint had been made to the US Federal Trade Commission (FTC) about RiceTec's description of its rice as "basmati", but the FTC took the view that this was a generic term and that consumers would not be deceived by the description "American basmati" (see Lightbourne, 2003; Subbiah, 2004, 554).

An example of patenting from an ex-situ collection maintained by a CGIAR institute involved the patenting of a gene from a strain of rice (*Oryza longistaminata*), originally from Mali. In the late 1970s *O longistaminata* was identified by a

[1] US Patent US5411736 A.

[2] US Patent 5,663,484.

[3] U.S. Patent No. 5,663,484, Reexamination Certificate C1 (4525th) (reissued Jan. 29, 2002).

researcher working in Cuttack North India, as being resistant to bacterial blight. In 1978, this resistant sample was taken to IRRI in Los Banos, Philippines for further investigation. Over a 15 year period, through conventional breeding IRRI researchers developed, a high-yielding, blight resistant strain of rice. A post-doctoral research fellow from the University of California at Davis, working at IRRI, was permitted with co-workers at Stanford University to map, sequence and clone the gene Xa21, which was identified as the genetic locus which contributed the resistance to blight. On 7th June 1995 the Regents of the University of California filed a patent application for "Nucleic acids, from *Oryza sativa*, which encode leucine-rich repeat polypeptides and enhance *Xanthomonas* resistance in plants." The patent was granted by the United States Patents and Trademark Office on 12 January 1999.[4] This patent generated some controversy because it was perceived to compromise IRRI's research efforts and those of its clients in the rice-producing regions of Asia. Bacterial blight is not a particular problem for US rice producers and a primary effect of the patent was to prevent the export of bacterial blight resistant rice, utilising the patent to the USA. This patent also raised the question of equitable compensation, at least for the traditional farmers of Mali who had conserved *O. longistaminata* (WIPO/UNEP, 2001, 13).

In 1995 and 2000 it was reported that University of Wisconsin scientists had patented and were exploiting patents on "brazzein" a protein extracted from the berries of *Pentadiplandra brazzeana* from Gabon. Natur Research Ingredients, Inc., a US corporation, was reported in late 2008 to have acquired the sole rights to manufacture and distribute brazzein from the University of Wisconsin at Madison (Micalizzi, 2017). This exploitation of Brazzein was cited as an instance of biopiracy to the UK Parliament's Select Committee on Environmental Audit in 1999 (UK Parliament, 1999) and is referred to as the classic exemplar of biopiracy (Brody, 2010, 51).

Another illustration of biopiracy influencing the international intellectual property environment is the so-called Basmati affair. This commenced when RiceTec, an American company based in Alvin, Texas, was granted a patent by the USPTO in September 1997 for "Basmati rice lines and grains".[5] The "novel rice lines" were described in the patent as "lines whose plants are semi-dwarf in stature, substantially photoperiod insensitive and high yielding" and which "produce rice grains having characteristics similar or superior to those of good quality basmati rice". In March 1998 an Indian NGO, the Research Foundation for Science, Technology and Ecology, petitioned India's Supreme Court to direct the government to challenge the patent, or to commence an action with the Dispute Settlement Body of the WTO. The Indian Government commenced an action in the USPTO in April 2000, challenging three of the patent claims (15–17). In response, RiceTec withdrew a number of its claims.

---

[4] U.S. patent 5,859,339.

[5] Patent 5,663,484 (USPTO).

Probably the most notorious example of agricultural biopiracy concerned a patent granted by the US Patent and Trademarks Office of a patent on April 13, 1999 for an invention relating to "a new field bean variety that produces distinctly coloured yellow seed which remain relatively unchanged by season."[6] The applicant was the president of a Colorado-based seed company, Pod-ners, which was reported to have written to all US importers of Mexican beans requiring the payment of a royalty of six cents per pound (Rattray, 2002). Pod-ners was reported to have brought infringement actions against two companies that were selling the Mexican yellow beans in the US. In January 2000, the Mexican government announced that it would challenge the US patent and on 20 December 2000 CIAT filed a formal request for re-examination of the patent claiming that the patent "would establish a precedent threatening public access to plant germplasm… held in trust by CIAT and research centers worldwide" (CIAT, 2008) The basis of the re-examination was that the patent failed to meet the statutory requirements of novelty and non-obviousness (See Nottenburg, 2009). CIAT argued that of its 260 bean samples with yellow seeds, six of the accessions were "substantially identical" to claims made in the patent. CIAT's patent challenge also asserted that the yellow bean was "misappropriated" from Mexico, and that this was in breach of Mexico's sovereign rights over its genetic resources. By way of a cross-claim, Pod-ners filed a request for a reissue of the patent on the basis that certain prior art had not been considered in the original application. In the re-examination the Examiner rejected the patentee's claims as obvious, explaining that the Enola plant and seed appeared to be genetically identical to the yellow Azufrado Peruano 87 bean. The USPTO's Board of Patent Appeals and Interferences upheld the rejection, concluding that the Examiner had established a prima facie case of obviousness which Pod-ners had failed to rebut. In 2009 Pod-ners failed in an appeal to the Federal Circuit Court.

In 2003, the Peruvian government identified several patents and patent applications relating to 'maca' (*Lepidium meyenii*), which had traditionally been cultivated in the Andes, including claims concerning therapeutic methods and uses of the plant (WIPO IGC, 2003). The Peruvian government expressed its concerns about the extent to which the patents and pending applications in the USA could prevent exports of maca extracts from Peru. Similarly, from 2001 the Japanese company Asahi Foods Co., Ltd. and an associated US company "Cupuacu International Inc" had obtained a number of patents on the extraction of lipids from the cupuaçu seeds. The pulp of cupuaçu (*Theobroma Grandiflorum*), which grows in the rainforests of Brazil, is used by traditional peoples to make fresh juice or as a sweetener for confectionary and as a medicament (See Matthews, 2011, 150–156).

Another example of "biopiracy" from *in situ* resources is the patenting of a gene isolated from *Streptomyces viridochromogenes* a micro-organism isolated from Cameroonian soil, which is responsible for the tolerance to glufosinate herbicides.[7]

---

[6] US Patent 5,894,079.

[7] US patent No. 5,276,268.

Despite the successful commercialisation of this chemical, no benefits had been shared with Cameroon (Mahop, 2006, 132).

A 2006 study by the Edmonds Institute, in cooperation with the African Centre for Biosafety, identified 36 instances of biopiracy, including the patenting of endophytes for improving fescues from North Africa Morocco and Tunisia and nematocidal fungi from Burkina Faso, as well as attributes of Ethiopian Teff (McGown, 2006).

## 3.4 Plant Variety Rights Protection

The 2006 African case studies by the Edmonds Institute included the utilization by American breeders of groundnut varieties from Malawi, Mozambique, Nigeria, Senegal and Sudan (McGown, 2006).

Concerns were raised in 1998 about plant breeder's rights (PBR) applications made in Australia by a number of agricultural research institutes in relation to a peavine and a lentil which had been bred from genetic stock obtained from ICARDA. A feature article in the *New Scientist* carried an accusation from a spokesperson from the South Asian Network on Food, Ecology and Culture which described the PBR applications as "blatant biopiracy" by "privatising seeds that belong to our farmers and selling them back to us". (Edwards & Anderson, 1998). CGIAR Chairman, Dr. Ismail Serageldin, called for a moratorium on the distribution of germplasm as "the strongest signal the CGIAR can send governments to ensure that … the materials in the CGIAR remain in the public domain" (CGIAR, 1998). To prevent a recurrence of this incident, the operating regulations of the Australian Plant Breeders Rights Office were amended to oblige applicants for PBRs in relation to varieties derived from germplasm obtained from CGIAR centres, to document that such applications were made with the permission of the relevant centre.

In November 1999, five traditional Peruvian varieties of yacon (Smallantus sonchifolius) an ancient Andean fruit held at the International Potato Center (CIP) in Peru, were distributed by the Peruvian Ministry of Agriculture to researchers in Japan. Yacon has a high fructose content with a high percentage of insulin and with antidiabetic properties. In 2000, Japanese researchers reported that the National Shikoku Agriculture Experiment Station had released the first commercial variety of yakon, "Sarada-Otome", on August 25, 2000 (Huaman, 2001). CIP's potato curator, Dr. Huaman expressed concern that Japanese researchers were not prepared to send germplasm of the new variety to be tested in Peruvian farmers' fields, thereby denying a source country of new derivatives of deposited germplasm (Huaman, 2001). CIP's Genetic Resources Policy Committee (GRPC), chaired by Dr. M. S. Swaminathan, concluded that CIP had no right to interfere in Peru's sovereign decision to send the germplasm to Japan and commended CIP for its proper management of its germplasm held "in-trust (Blakeney, 2001).

Responding to concerns about the impact of intellectual property rights upon the operation of the CGIAR, it commissioned a report on the use of proprietary technologies by CGIAR Centres by the International Service for National Agriculture Research (ISNAR), which operated as its legal advisory body (Cohen, Falconi, Komen, & Blakeney, 1998). The report noted the burgeoning use of proprietary technologies by the centres and recommended that they undertake audits of their intellectual property management policies. These cases led to an intense discussion within the CGIAR of the approach to be taken within the organization to intellectual property rights. Some CGIAR Centres perceive that CGIAR-generated intellectual property might be used as a bargaining chip, to be traded for biological tools patented by the private sector. For example the Policy on Intellectual Property of the International Maize and Wheat Improvement Center (CIMMYT) envisages that intellectual property protection may be sought "to facilitate the negotiation and conclusion of agreements for access to proprietary technologies of use to CIMMYT's research and in furtherance of its mission.[8] This proprietisation of public sector agriculture research is questioned, particularly by those NGO's opposed to patenting in the life sciences (see Blakeney, 2000).

## 3.5 Traditional Knowledge and Identification of Useful Genetic Resources

The traditional knowledge of indigenous peoples and farmers has played an important role in identifying biological resources worthy of commercial exploitation. For example, the search for new pharmaceuticals from naturally occurring biological material has been guided by ethnobiological data (See McChesney, 1996; ten Kate & Laird, 2000) In a number of the "biopiracy" examples above, the knowledge of local communities, traditional and indigenous peoples was utilised to identify useful germplasm. The utilisation of this knowledge in identifying biologically active substances has saved bio-prospectors the considerable amounts of money they would otherwise have expended in screening substances plucked at random. Thus, "biopiracy" often involves both the unauthorised access to biological materials and the unauthorised exploitation of the knowledge used to identify those materials as useful. The close relationship between identifying useful genetic resources and traditional knowledge is emphasized by Sharma et al. (2018) who point out that about two-third of Indian population relies on indigenous knowledge of biological resources and that more than 7500 species of plants are utilized for the traditional purposes in India. They then list 17 specific cases of the unauthorized patenting of Indian biological resources used and conserved by traditional communities.

Examples of traditional knowledge with and agricultural application include: "mental inventories of local biological resources, animal breeds, and local plant, crop, and tree species" as well as plants which are indicators of soil salinity, seed

[8] CIMMYT, *Policy on Intellectual Property*, Article III.4.v, available at www.cimmyt.org/resources/obtaining/seed/ip_policy/htm/ip-policy.htm, accessed 18 October 2019.

treatment and storage methods and tools used for planting and harvesting (Hansen, 2007). A similarly significant contribution has been made by the knowledge of indigenous peoples and farmers in the development of new crop types and biodiversity conservation. These groups have been an important agency in the conservation of plant genetic resources and the transmission of these resources to seed companies, plant breeders and research institutions. They have not typically been paid for the value they have delivered, whereas breeders and seed companies have resorted to intellectual property rights to recover their development expenditures. The economic value of biological diversity conserved by traditional farmers for agriculture is difficult to quantify and it has been suggested that "the value of farmers' varieties is not directly dependent on their current use in conventional breeding, since the gene flow from landraces to privately marketed cultivars of major crops is very modest" because "conventional breeding increasingly focuses on crosses among elite materials from the breeders own collections and advanced lines developed in public institutions." (Wright, 1998). On the other hand, those collections and advanced breeding lines are often derived from germplasm contributed by traditional groups.

An example of the patenting of genetic resources identified with the assistance of traditional peoples, concerns Camu camu (*Myrciaria dubia*) a plant with very high levels of ascorbic acid (vitamin C), used by traditional peoples in the Peruvian Amazon. In October 2005 Peru notified the World Trade Organization of "potential biopiracy" arising from a series of international patents and patent applications, principally published under the Patent Cooperation Treaty (PCT) and by Japanese Patent Office for skin preparations, cosmetics and food additives utilizing camu camu (Peru, 2005). This notification was also communicated to the World Intellectual Property Organization (Peru, 2006).

In 2006 Brazil in a submission expressed its concern to the WIPO Standing Committee on Trademarks, Industrial Designs, and Geographical Indications about a number of patents and trademarks on its plants (Brazil, 2006). One of these was a patent on Açai (*Euterpe oleracea*) a fruit that had been traditionally used in Brazil as a food and medicine obtained by a US Corporation, Mary Kay Inc. The same corporation had obtained a US patent[9] concerning the processing of the Kakadu plum (*Terminalia ferdinandiana*), a traditional food and medicine source for Aboriginal Peoples in Northern Australia (Gorman, Griffiths, & Whitehead, 2006). On January 19, 2007 Mary Kay Inc. applied under the Patent Cooperation Treaty to extend this patent to more than 100 countries.[10] The patent application entered the national phase in Australia on 22 July 2008.

The Examination Report issued by IP Australia, stated reported its finding that the "Aborigines have been using the Kakadu plum extract for around 40,000 years

[9] US Patent 7175862.

[10] WO/2007/084998.

as a food source and a healing agent."[11] On 12 October 2011 the Australian application was withdrawn, although the granted patents, as well as patent applications remain on foot in a number of countries (Robinson, 2010).

The bioprospecting issue which has been raised in relation to this case concerns the source of the Kakadu plum used by Mary Kay, Inc. It has been pointed out that if it was obtained by a commercial supplier, there would have been no obligation to share benefits with Indigenous communities under local or international legislation (Holcombe & Janke, 2012 at 309–11).

Discussed below are measures and proposed measures for securing consent to access traditional knowledge and the associated biological resources identified as useful and measures to secure the equitable sharing of commercial benefits with farmers and traditional communities.

## 3.6 Convention on Biological Diversity (CBD)

Most of the biodiverse countries are located in tropical and sub-tropical areas and most of them, from an economic perspective are developing or least developed countries (LDCs). In other words, their biological wealth has not been translated into economic wealth. One of the reasons for this is the absence of a binding global legal regime which obliges the exploiters of genetic resources to seek the consent of source countries.

The Rio Earth Summit, which was convened in June 1992, promulgated the CBD which represented an attempt to establish an international programme for the conservation and utilization of the world's biological resources. "The single most divisive issue in the negotiations was the relationship between intellectual property rights and access to genetic resources" (Chandler, 1993, 161), in particular the conditions for access and benefit sharing. Article 1 of the CBD envisages "appropriate access to genetic resources" and "the fair and equitable sharing of benefits arising out of the utilization of genetic resources". "Genetic resources" are defined in Art.2 as meaning "genetic material of actual or potential value". The term "genetic material" is then defined in Art.2 to mean "any material of plant, animal, microbiological or other origin containing functional units of heredity". Thus, the CBD applies to seeds and cuttings and DNA extracted from a plant, such as a chromosome, gene, plasmid or any part of these such as the promoter part of a gene (See Glowka, 1998, 4).

Article 9 deals with "the conservation of components of biological diversity outside their natural habitats", for example, in germplasm and seed banks, botanical gardens, museums, laboratories and agricultural research institutions. This article calls for national legislation to provide for the acquisition, conservation, storage and

[11] The Examination Report is available at http://pericles.ipaustralia.gov.au/ols/auspat/applicationDetails.do?applicationNo=2007205838, accessed 18 October 2019.

management of these *ex situ* collections. Article 15(3) provides that the access and benefit-sharing provisions of the CBD do not apply to the genetic resources of a country which were collected prior to the entry of the CBD into force in that country. Thus, a country with a pre-existing collection of genetic material has the sovereign right to control access to that collection, but has no legal right to insist upon a share of any benefits derived from the use of that collection (Yusuf, 1994).

Article 15(1) of the CBD affirms "the sovereign rights of States over their natural resources" and provides that "the authority to determine access to genetic resources rests with the national governments and is subject to national legislation". Article 15(4) of the CBD envisages that where access is granted it will be subject to mutually agreed terms. Article 15(7) requires each Contracting Party to "take legislative, administrative or policy measures, as appropriate" and in accordance with a number of specified provisions of the Convention, "with the aim of sharing in a fair and equitable way, the results of research and development and the benefits arising from the commercial and other utilization of genetic resources with the Contracting Party providing such resources".

Complementary to the equitable sharing of benefits, the CBD provides for the access of developing country signatories to technologies which may result from the utilisation of the genetic resources which they may provide. Article 16(1) recites the importance of access to biotechnologies to attain the objectives of the CBD and Art 16(2) provides for the access to technologies by developing countries on "fair and equitable terms, including on concessional and preferential terms". Article.19(1) requires parties to take appropriate measures to "provide for the effective participation in biotechnological research activities by those Contracting Parties, especially developing countries, which provide the genetic resources for such research". Article 19(2) requires parties to "take all practicable measures to promote and advance priority access on a fair and equitable basis,…,especially developing countries, to the results and benefits arising from biotechnologies based upon genetic resources provided by those Contacting Parties" on mutually agreed terms.

The Rio Declaration in Principle 22 stated that "Indigenous peoples and their communities...have a vital role in environmental management and development because of their knowledge and traditional practices". The Preamble to the CBD recognised the

> …close and traditional dependence of many Indigenous and local communities embodying traditional lifestyles on biological resources, and the desirability of sharing equitably arising from the use of traditional knowledge, innovations and practices relevant to the conservation of biological diversity and sustainable use of its components.

Article 8(j) of the Convention required each signatory

> …subject to its national legislation, respect, preserve and maintain knowledge, innovations and practices of indigenous and local communities embodying traditional lifestyles relevant for the conservation and sustainable use of biological diversity and promote their wider application with the approval and involvement of the holders of such knowledge, innovations and practices and encourage the equitable sharing of the benefits arising from the utilization of such knowledge, innovations and practices.

The provisions of Art.8(j) require implementation through national legislation. It is expressed to be subject to national legislation, in order to preserve legislation on this subject which predates the CBD.

The discussion, in the context of the CBD, of the intellectual property rights of traditional and local communities has not tended to focus upon the rights of traditional farming communities. This subject has been taken up as an aspect of the International Treaty on Plant Genetic Resources for Food and Agriculture, which is discussed below.

## 3.7 The Nagoya Protocol

The CBD did not set out how access and benefit-sharing (ABS), envisaged in Arts 15, 16, 19(2) and 8j would be implemented. At the conference of the parties (COP) of the CBD in October 2001, an Ad Hoc Open-Ended Working Group on ABS was established and at its first meeting in Bonn, it developed the Bonn Guidelines on Access to Genetic Resources and Fair and Equitable Benefit Sharing which was adopted by the seventh COP on a non-binding, voluntary basis.[12] The contribution of traditional peoples referred to in Art. 8j of the CBD was decision taken into account by further sessions of the Working Group and in 2010 the COP adopted the Nagoya Protocol on Access to Genetic Resources and the Fair and Equitable Sharing of Benefits Arising from their Utilization to the Convention on Biological Diversity.[13] Article 6 of the Protocol reiterated the CBD's recognition of country's sovereign rights over natural resources and that access to genetic resources be subject to the prior informed consent (PIC) and on mutually agreed terms (MAT). Article 5 of the Protocol provided that the benefits arising from the utilization of genetic resources "as well as subsequent applications and commercialisation" are to be shared with the provider of those resources in a fair and equitable way. Article 7 provides that "in accordance with domestic law, each Party shall take measures, as appropriate" with the aim of ensuring that TK associated with genetic resources that is held by indigenous and local communities is accessed with the prior and informed consent or approval and involvement of these indigenous and local communities, on the basis of mutually agreed terms. Article 12.1 of the Protocol requires Parties in implementing their obligations "in accordance with domestic law" take into consideration indigenous and local communities' customary laws, community protocols and procedures, as applicable, to TK associated with genetic resources. The Nagoya Protocol entered into force on 12 October 2014, 90 days after the deposit of the fiftieth instrument of ratification.

---

[12] 'Bonn Guidelines on Access to Genetic Resources and Fair and Equitable Sharing of the Benefits Arising out of their Utilization' in Report of the Sixth Meeting of the Conference of the Parties to the Convention on Biological Diversity, UN Doc. UNEP/CBD/COP/6/20 (2002).

[13] UNEP/CBD/COP/10/L.43/Rev.129 October 2010.

In June 2015 the African Union adopted the *African Union Practical Guidelines for the Coordinated Implementation of the Nagoya Protocol in Africa* which provide" a practical step by step guidance for the implementation of the Protocol and for an ABS system at national and regional levels." An important feature of the Guidelines is their stress on the importance of identifying and involve all stakeholders, ranging from private and communal traditional knowledge holders and/or land owners who have legal rights to provide access to genetic resources, local researchers and business people involved in bio-prospecting either as intermediaries or end users, and various government authorities tasked with regulating specific habitats (e.g. protected areas) or sets of resources (e.g. marine resources) or legal aspects.

## 3.8 International Treaty on Plant Genetic Resources for Food and Agriculture

The specific issue of the biopiracy of genetic resources from the international agricultural research centres of the CGIAR was sought to be dealt with by the 2001 International Treaty on Plant Genetic Resources for Food and Agriculture, which entered into force on 29 June, 2004. Article 10.2 of the Treaty contains the agreement of the Contracting Parties to "establish a multilateral system, which is efficient, effective and transparent, both to facilitate access to Plant Genetic Resources for Food and Agriculture (PGRFA) and to share, in a fair and equitable way, the benefits arising from the utilisation of these resources, on a complementary and mutually reinforcing basis". The PGRFA to which the Multilateral System applies are some 35 crops and 29 forages which are listed in Annex I and other contributions by resource holders (Art 11(2)). The collections of the CGIAR are expressly included in the Multilateral System (Art. 11(5)). Access to PGRFA of such crops and forages is to be provided free or at a minimal cost.

The Treaty attempts to create an international genetic resources commons by seeking to limit the propertisation of the categories of crops and forages to which it applies (Halewood & Nnadozie, 2008, 115).

The International Treaty in Article 12.3 provides that facilitated access to PGFRA is to be provided under Material Transfer Agreement on condition (d) that the recipients "shall not claim any intellectual property or other rights that limit the facilitated access" to PGFRA, or their "genetic parts or components", in the form received from the Multilateral System. This, of course, does not prevent intellectual property rights being claimed in relation to germplasm which is modified by the recipient. A problematic issue is the extent of modification which must occur before it can be said that the form in which the germplasm was received has changed.

A Standard Material Transfer Agreement (SMTA) to be used for accessions of material falling within the International Treaty was finalised in 2006 (FAO, 2006). The parties to the SMTA agree in Article 4.3 that the Governing Body of the Treaty and its Multilateral System (ie the Food and Agricultural Organization of the United

Nations (FAO)) is identified as the third party beneficiary under the SMTA. Including the FAO as the third party beneficiary puts it in a position to enforce the SMTA. The limited financial resources for legal enforcement actions of many of the institutes which will be supplying genetic resources under SMTAs means sets up the FAO as a more likely litigant. However, Article 4.5 preserves the rights of the provider and the recipient from exercising their rights under the SMTA. Although the SMTA seeks to construct a legal basis for the enforcement of rights in relation to germplasm and other materials supplied under its terms, the greater likelihood is that the SMTA will be enforced as a moral obligation. Also, recipients who do not abide by the terms of a SMTA are likely to be excluded from the receipt of any further material under the multilateral system.

Article 5 of the SMTA provides that in the case of transfers from CGIAR Centres these will be subject to the Agreement between the FAO and the Centres under which trusteeship of their collections is conferred on the FAO. Article 5 (d) provides that access to PGRFA protected by intellectual and other property rights shall be consistent with relevant international agreements, and with relevant national laws, but under Art. 6.2 the recipient agrees not to claim any intellectual property or other rights that limit the facilitated access to the material provided under the SMTA or its genetic parts or components, in the form received from the Multilateral System. This terminology leaves it open for recipients to obtain intellectual property rights in modified derivatives.

Where a recipient obtains intellectual property rights on any products developed from the material supplied under a SMTA, or its components and assigns such intellectual property rights to a third party, Art. 6.10 requires that the recipient shall transfer the benefit-sharing obligations of the SMTA, set out in Art. 6.7 to that third party. Under Art. 6.1 of the SMTA the recipient undertakes that the material shall be used or conserved only for the purposes of research, breeding and training for food and agriculture. Such purposes shall not include chemical, pharmaceutical and/or other non-food/feed industrial uses.

Article 13.1 of the International Treaty, recognises that benefits accruing from facilitated access to PGFRA shall be shared fairly and equitably under this Article. Article 13.2 envisages that this sharing of benefits includes the exchange of technical information, access to technology, capacity building and the sharing of monetary benefits from commercialisation.

## 3.9 Farmer's Rights Under the International Treaty on PGRFA.

The concept of Farmers' Rights was developed as "a counterbalance to intellectual property rights (FAO, 1994)." Farmers' rights were intended to promote a more equitable relation between the providers and users of germplasm by creating a basis for farmers to share in the benefits derived from the germplasm which they had developed and conserved over time (see Glowka, 1998, 20). Under Art. 5.1(c) of the International Treaty the Contracting Parties agreed, subject to national legislation, to promote or support, as appropriate, farmers and local communities' efforts to manage and conserve on-farm their plant genetic resources for food and agriculture and in Art. 5.1(d) to promote *in situ* conservation of wild crop relatives and wild plants for food production, by supporting, *inter alia*, the efforts of indigenous and local communities.

Article 9.2 of the International Treaty envisaged that "the responsibility for realizing Farmers' Rights…rests with national governments" and that national legislation should include measures relating to:

(a) protection of traditional knowledge relevant to plant genetic resources for food and agriculture;
(b) the right to equitably participate in sharing benefits arising from the utilization of plant genetic resources for food and agriculture;
(c) the right to participate in making decisions, at the national level, on matters related to the conservation and sustainable use of plant genetic resources for food and agriculture.

An assumption of Art.9 was that the landraces used by traditional farmers are a dynamic genetic reservoir for the development of new varieties and for the transmission of desirable genetic traits. The traditional knowledge of local and indigenous communities which permits the identification of useful plants is similarly perceived. The diversity of landraces and the associated information on their specific qualities contribute invaluable information to formal breeding processes.

At the periodic meetings of the Governing Body of the International Treaty member states have differed on the way in which farmers' rights might be implemented. On the one hand the industrialised agricultural states such as Australia, Canada and the USA have argued that the implementation of farmers' tights is a matter for national legislation.[14] On the other hand, calls have been made by developing countries, such as the Africa group that the Governing Body of take measures to support the Contracting Parties technically and financially in:

(a) building farmers' capacity to participate in decision-making regarding their rights to save, use, exchange and sell farm-saved seed and propagating material;
(b) facilitating access to relevant information regarding the laws and policies pertaining to farmers' rights to save, use, exchange and sell farm-saved seed;
(c) ensuring effective participation of farmers in such decision-making;
(d) raising awareness among farmers, policy-makers and other relevant groups;
(e) establishing legal support for informal seed systems;
(f) mainstreaming Farmers' Rights to save, use, exchange and sell farm-saved seed in legal and policy frameworks;

---

[14] Eg Seventh session of the Governing Bodies held in Kigali, 30 October to 3 November 2017, International Treaty Doc., IT/GB-7/17/L12 Rev.1

(g) up-scaling and institutionalizing successful local activities aimed at strengthening informal seed systems, including NGO-led activities, to the national level;
(h) harmonizing seed regulation in the region to protect Farmers' Rights.[15]

In relation to the realization of measures to protect traditional knowledge the Governing Body was requested by the Africa Group to support the Contracting Parties in:

(a) establishing measures to recognize traditional knowledge and facilitate its use;
(b) establishing measures to ensure that traditional knowledge, as well as the systems that generate such knowledge, are respected and promoted;
(c) facilitating documentation of traditional knowledge;
(d) making use of media to ensure publicity for traditional knowledge;
(e) building capacity for documenting and using traditional knowledge;
(f) establishing measures for scaling up documentation and use of traditional knowledge;
(g) developing and implementing legal provisions on traditional knowledge;
(h) supporting on-farm conservation activities by farmers.[16]

World Trade Organization (WTO) Agreement on Trade Related Aspects of Intellectual Property Rights (TRIPS).

Facilitating the biopiracy of genetic resources has been the establishment of a global patent regime pursuant to the TRIPS Agreement. Art. 27 of TRIPS requires that patents be available in all fields of technology This will include biotechnology and is obviously in tension with the objectives of the CBD and the International Treaty. It has been suggested that the TRIPS Agreement should be amended so as to require, or to enable, WTO Members to require that patent applicants disclose, as a condition to patentability: (a) the source of any genetic material used in a claimed invention; (b) any related traditional knowledge used in the invention; (c) evidence of prior informed consent from the competent authority in the country of origin of the genetic material; and (d) evidence of fair and equitable benefit sharing and that such provisions could be incorporated into the TRIPS Agreement by amendment.[17]

## 3.10 World Intellectual Property Organization (WIPO) and Genetic Resources

In September 1999, the delegation of Colombia proposed the introduction into the Patent Law Treaty, then under negotiation, that an article be inserted which provided that:

---

[15] Global Consultation Conference on Farmers' Rights held in Addis Ababa in November 2010, reproduced in IT/GB-4/11/Circ.1, Annex 1, para.11.

[16] Ibid., para 12.

[17] WTO Doc. IP/C/W/228, IP/C/M/32, para. 128, IP/C/M/33, para. 121 (Brazil).

1. All industrial protection shall guarantee the protection of the country's biological and genetic heritage. Consequently, the grant of patents or registrations that relate to elements of that heritage shall be subject to their having been acquired made legally.
2. Every document shall specify the registration number of the contract affording access to genetic resources and a copy thereof whereby the products or processes for which protection is sought have been manufactured or developed from genetic resources, or products thereof, of which one of the member countries is the country of origin.

The Diplomatic Conference, which commenced on 11 May, 2000, became bogged down on the question of obliging the identification of source countries in biotechnological patent applications. To facilitate progress on the procedural aspects, the source country question was referred to an expert group for further consideration. At the WIPO General Assembly in 2000 the Member States agreed the establishment of an Intergovernmental Committee on Intellectual Property and Genetic Resources, Traditional Knowledge and Folklore (IGC). Three interrelated themes were identified to inform the deliberations of the Committee: intellectual property issues that arise in the context of (i) access to genetic resources and benefit sharing; (ii) protection of traditional knowledge, whether or not associated with those resources; and (iii) the protection of expressions of folklore (WIPO, 2000).

The early sessions of the IGC were concerned with the formulation of model guidelines and intellectual property clauses for contractual agreements on access to genetic resources and benefit-sharing (eg WIPO, IGC, 2001). At the same time the IGC has concerned itself with formulating treaties for the protection of traditional knowledge and traditional cultural expressions. This has been a long drawn out process, largely attributable to conflicts between bioprospecting and source countries, as well as to tensions between traditional and dominant communities (Blakeney, 2016).

There is not yet a draft treaty text on the protection of genetic resources, but a "Consolidated Document Relating to Intellectual Property and Genetic Resources" (WIPO, IGC, 2019). The negotiations are very far from conclusion. There is not yet even an agreed preamble, nor agreed definitions of terms. In any event, for a global regime based upon this text to be effective, national legislation will have to sanction the use of genetic resources obtained without informed consent or without benefit-sharing arrangements.

The mandate of the IGC for 2020/2021 is to:

> …continue to expedite its work, with the objective of finalizing an agreement on an international legal instrument(s), without prejudging the nature of outcome(s), relating to intellectual property which will ensure the balanced and effective protection of genetic resources (GRs), traditional knowledge (TK) and traditional cultural expressions (TCEs).[18]

[18] https://www.wipo.int/export/sites/www/tk/en/igc/pdf/igc_mandate_2020-2021.pdf, accessed 22 October 2019.

## 3.11 Conclusion

In the absence of an effective international legal regime to regulate biopiracy, a second-best solution is for source countries to regulate access to their genetic resources. Among the pioneering legislation in this regard is the Indian Biodiversity Act of 2002 which provides that "no person shall apply for any intellectual property right … in or outside India for any invention based on any research or information on a biological resource obtained from India without obtaining the previous approval of the National Biodiversity Authority before making such application, provided that if a person applies for a patent, permission of the National Biodiversity Authority may be obtained after the acceptance of the patent but before the sealing of the patent by the patent authority concerned".[19]

This legislation seems to accord with world's best practice of nesting bioprospecting within the broader environmental legal framework which will allow a greater degree of certainty "in the relationship between overlapping laws and policies." (Cabrera et al., 2012. 18). Similarly, South Africa has enacted its National Environmental Management: Biodiversity Act, 2004, which regulates bioprospecting, within the framework of the National Environmental Management Act, 1998.

## References

Blakeney, M. (1997). Protection of traditional medical knowledge of indigenous peoples. *European Intellectual Property Review, 19*, 298–302.

Blakeney, M. (1998). Intellectual property rights in the genetic resources of international agricultural research institutes- some recent problems. *Biosci Law Rev, 1*, 3–11.

Blakeney, M. (2000). Protection of traditional knowledge under intellectual property law. *European Intellectual Property Review, 22*, 251–261.

Blakeney, M. (2001) 'Intellectual property aspects of traditional agricultural knowledge', in IP in Blakeney, M. and Drahos, P. (eds.) Biodiversity and agriculture: Regulating the biosphere, London, Sweet & Maxwell, 29–52.

Blakeney, M. (2004). Bioprospecting and biopiracy. In B. Ong (Ed.), *Intellectual property and biological resources* (pp. 393–424). Singapore, Singapore: Marshall Cavendish.

Blakeney, M. (2016). The negotiations in WIPO for international conventions on traditional knowledge and traditional cultural expressions. In J. C. Lai & A. M. Dominicé (Eds.), *Intellectual property and access to im/material goods*. Cheltenham, UK: Edward Elgar.

Blakeney, M. (2019). Remedying the misappropriation of genetic resources. In H. B. Singh, C. Keswani, & S. P. Singh (Eds.), *Intellectual property issues in microbiology* (pp. 147–162). Singapore, Singapore: Springer.

Brazil (2006) WIPO Doc, SCT/16/9.

Brody, B. A. (2010). Intellectual property, state sovereignty, and biotechnology. *Kennedy Institute of Ethics Journal, 20*(1), 50–73.

Cabrera, J. et al. (2012). *Survey of future just biodiversity laws & policies*. Available at https://www.worldfuturecouncil.org/file/2016/01/WFC_CISDL_2012_Survey_of_Future_Just__Biodiversity_Policies_and_Laws.pdf. Accessed 22 Oct 2019.

[19] Section 6(1), Indian Biodiversity Act of 2002.

CGIAR. (1998). *CGIAR urges halt to granting of intellectual property rights for designated plant germplasm*. Press Release, February 11, quoted in Blakeney, (2001).
Chandler, M. (1993). The biodiversity convention: Selected issues of interest to the international lawyer. *Colorado Journal of International Law and Policy, 4*(1), 140–175.
Chen, J. (2000). *Diversity and deadlock: Transcending conventional wisdom on the relationship between biological diversity and intellectual property*. CASRIP Publication Series: Rethinking Int'l Intellectual Property, No.6 (pp. 167–217).
Chen, J. (2005). There's no such thing as biopiracy … And It's a good thing too. *McGeorge Law Review, 36*, 1–35.
CIAT. (2008, May 2). *US Patent Office rejects US company's patent protection for bean commonly grown by Latin American Farmers*. Press Release. Available at www.cgiar.org. Accessed 18 Oct 2019.
Cohen, J., Falconi, C., Komen, J., & Blakeney, M. (1998). *The use of proprietary biotechnology research inputs at selected CGIAR centres*. The Hague, The Netherlands: CGIAR.
De Pace, C., et al. (2011). Deployment of either a whole or dissected wild nuclear genome into the wheat gene pool meets the breeding challenges posed by the sustainable farming systems. *Plant Genetic Resources: Characterization and Utilization, 9*(2), 352–356.
Dias, J. S. (2015). Biodiversity and plant breeding as tools for harmony between modern agriculture production and the environment. In M. Caliskan, et al. (Eds.), *Molecular approaches to genetic diversity* (pp. 3–44). Rijeka, Croatia: Intexch.
Edwards, R., & Anderson, I. (1998). Seeds of wrath. *New Scientist, 157*(2121), 14–15.
Efferth, T. (2019). Biopiracy of medicinal plants: Finding fair solutions for the use of natural resources. *Phytomedicine, 53*, 294–295.
ETC Group. (2005). *ETC Group Web Site*. Available at: http://www.etcgroup.org/text/txt_key_defs.asp. Accessed 14 Oct 2019.
FAO. (1994). *The State of the World's Plant Genetic Resources for Food and Agriculture*. Rome: FAO.
FAO. (2006). *Report of the First Session of the Governing Body of the International Treaty on Plant Genetic Resources for Food and Agriculture* (IT/GB-1/06/Report). Madrid, Spain.
Fowler, C., & Hodgkin, T. (2004). Plant genetic resources for food and agriculture: Assessing global availability. *Annual Review of Environmental Resources, 29*, 10.1–10.37.
Glowka, L. (1998). *A guide to designing legal frameworks to determine access to genetic resources*. Gland, Switzerland: IUCN.
Gorman, J. T., Griffiths, A. D., & Whitehead, P. J. (2006). An analysis of the use of plant products for commerce in remote aboriginal communities of northern Australia. *Economic Botany, 60*(4), 362–373.
Hajjar, R., & Hodgkin, T. (2007). The use of wild relatives in crop improvement: A survey of developments over the last 20 years. *Euphytica, 156*, 1–13.
Halewood, M., & Nnadozie, K. (2008). Giving priority to the commons: The international treaty on plant genetic resources for food and agriculture. In G. Tansey & T. Rajotte (Eds.), *The future control of food. A guide to international negotiations and rules on intellectual property, biodiversity and food security* (pp. 115–140). London: Earthscan.
Hansen, S. A., & Van Fleet, J. W. (2007). Issues and options for traditional knowledge holders in protecting their intellectual property. In A. Krattiger, R. T. Mahoney, L. Nelsen, et al. (Eds.), *Intellectual property management in health and agricultural innovation: A handbook of best practices* (pp. 1523–1538). Oxford: MIHR, and PIPRA.
Heald, P. J. (2003). The rhetoric of biopiracy. *Cardozo Journal of International and Comparative Law, 11*, 519.
Holcombe, S., & Janke, T. (2012). Patenting the Kakadu Plum and the Marjarla tree: Biodiscovery, intellectual property and indigenous knowledge. In M. Rimmer & A. McLennan (Eds.), *Intellectual property and emerging technologies: The new biology* (pp. 293–319). Cheltenham, UK: Edward Elgar.

Huaman, Z. (2001). *Unethical distribution to Japan of Yacon held in trust by CIP*, quoted in Blakeney, 2001.

Kloppenburg, J., & Kleinman, D. (1988). Seeds of controversy: National property versus common heritage. In J. Kloppenburg (Ed.), *Seeds and sovereignty: The use and control of plant genetic resources* (pp. 1–15). Raleigh: Duke University Press.

Lightbourne, M. (2003). Of rice and men. An attempt to assess the basmati affair. *The Journal of World Intellectual Property, 6*(6), 875–894.

Mahop, M. T. (2006). *Community rights and biodiversity regulations: Lessons from Cameroon and South Africa*. PhD thesis, Queen Mary, University of London.

Matthews, D. (2011). *Intellectual property, human rights and development. The role of NGOs and social movements*. Cheltenham, UK: Edward Elgar.

Maxted, N., & Kell, S. (2009). *Establishment of a global network for the in situ conservation of crop wild relatives: Status and needs* (Background study paper No. 39). Commission on Genetic Resources for Food and Agriculture, FAO, Rome.

Maxted, N., et al. (2012). Toward the systematic conservation of global crop wild relative diversity. *Crop Science, 52*(2), 774–785.

McChesney, G. (1996). Biological diversity, chemical diversity and the search for new pharmaceuticals. In M. Balick, E. Elisabetsky, & S. Laird (Eds.), *Medicinal resources of the tropical forest: Biodiversity and its importance to human health* (pp. 11–18). New York: University of Columbia Press.

McGown, J. (2006). *Out of Africa: Mysteries of access and benefit sharing*. Washington, DC: Edmonds Institute and African Centre for Biosafety.

McNeely, J. A. (2001). Biodiversity and agricultural development: The crucial institutional issues. In D. R. Lee & C. B. Barrett (Eds.), *Tradeoffs or synergies? Agricultural intensification, economic development and the environment* (pp. 399–412). Wallingford, UK: CABI.

Mittermeier, R. A. P., Robles. G., & Mittermeier, C. G. (Eds). (1989). *Megadiversity: earth's biologically wealthiest nations*. Mexico City: Cemex.

Micalizzi, J. (2017). Misappropriation of genetic resources in Africa. A study of: *Pentadiplandra Brazzeana, Impatiens Usambarensis*, and *Combretum Micranthum*. *Journal of Law, Technology & the Internet, 8*, 1–29.

Nottenburg, C. (2009). The Enola Bean Patent Controversy. *Harvest Choice*. Available at http://harvestchoice.org/sites/default/files/downloads/publications/Nottenburg2009HarvestChoice%2D%2DEnolaBeanControversy-5S.pdf. Accessed 17 Oct 2019.

Peru. (2005, May 30). *The patent system and the fight against biopiracy – The Peruvian experience.* WTO Document: WIPO/GRTKF/IC/8/12..

Peru. (2006, March 15). Analysis of Potential Cases of Biopiracy, WIPO Doc. WIPO/GRTKF/IC/9/10.

Plant Variety Protection, International Agricultural Research and Exchange of Germplasm: Blakeney, M. (2000). Legal aspects of the new sui generis regimes. In W. Padolina (Ed.), *Plant variety protection of Rice in developing countries* (Impacts on Research and Development) (pp. 9–22). Los Banos, Philippines: IRRI.

Prescott-Allen, R., & Prescott Allen, C. (1988). *Genes from the wild: Using wild genetic resources for food and raw materials* (2nd ed.). London: Earthscan Publications.

Rattray, G. N. (2002). The Enola bean patent controversy: Biopiracy, novelty and fish-and-chips. *Duke Law and Technology Review, 1*, 1–8.

Robinson, D. F. (2010). *Confronting biopiracy: Challenges, cases and international debates*. London: Earthscan.

Sharma, B., Maurya, S. S., & Brahmacharimayum, B. J. (2018). India's fight against agricultural and medicinal plants' biopiracy: Its implications on food security, traditional rights and knowledge degradation. *International Journal of Agriculture, Environment and Biotechnology, 11*(6), 881–887.

Shiva, V. (2000). *Stolen harvest: The hijacking of the global food supply*. London: Zed Books.

Shiva, V. (2001). *Protect or plunder? Understanding intellectual property right*. London: Zed Books.
Shiva, V. (2013). *The neem tree – a case history of biopiracy*. Available at http://www.twnside.org.sg/title/pir-ch.htm. Accessed 30 April 2020.
Shiva, V., & Holla-Bhar, R. (1996). Piracy by patent: The case of the neem tree. In G. Mander & E. Goldsmith (Eds.), *The case against the global economy* (pp. 146–159). San Francisco: Sierra Club Books.
Srivastava, M., (2011). *Bio piracy of medicinal plants & practices: Sacrilege of aboriginal Indian traditional knowledge*. Available at https://ssrn.com/abstract=2149878. Accessed 16 Oct 2019.
Subbiah, S. (2004). Reaping what they sow: The basmati rice controversy and strategies for protecting traditional knowledge. *Boston College. International and Comparative Law Review, 27*, 529–559.
ten Kate, K., & Laird, S. A. (2000). *The commercial use of biodiversity: Access to genetic resources and benefit*. London: Earthscan.
UK Parliament. (1999). *Appendix 7 to the minutes of evidence of the select committee on environmental audit, 1999*. London: UK Parliament.
UNEP. (2000, April 20). *Progress report on the implementation of the programmes of work-information on marine and coastal genetic resources including bioprospecting*. UNEP/CBD/COP/5/INF/7.
WIPO. (2000, August 25). *Matters concerning intellectual property genetic resources traditional knowledge and folklore*. WIPO Doc, WO/GA/26/6.
WIPO, IGC. (2001, September 10). *WIPO Doc*. WIPO/GRTKF/IC/2/3.
WIPO, IGC. (2003). *Patents referring to Lepidium meyenii (maca): responses of Peru*. Document submitted by the Delegation of Peru, WIPO/GRTKF/IC/5/13.
WIPO, IGC. (2019, April 9). *Consolidated Document Relating to Intellectual Property and Genetic Resources, Rev. 2*. WIPO Doc. WIPO/GRTKF/40/6.
WIPO/UNEP. (2001). *The role of intellectual property rights in the sharing of benefits arising from the use of biological resources and associated traditional knowledge* (Selected case studies). Geneva, Switzerland: WIPO.
Wright, B. D. (1998). Intellectual property and farmers' rights. In R. Evenson, D. Gollin, & V. Santaniello (Eds.), *Agricultural values of plant genetic resources* (p. 228). Wallingford, UK: CABI.
Yusuf, A. A. (1994). International law and sustainable development: The convention on biological diversity. In A. A. Yusuf (Ed.), *African yearbook of international law* (Vol. 2, pp. 109–137). The Hague, The Netherlands\Boston\London: Kluwer.

# Chapter 4
# Local Knowledge and Climate Change Adaptation

**Michael Blakeney**

**Abstract** This chapter looks at the impact of climate change upon agriculture and the concept of climate change adaptation and the role that local knowledge can play in that adaptation. It looks at issues of governance and seed conservation. The cultivation of neglected and underutilized species is examined together with the role of cultural factors in the adoption of those species. It concludes with an examination of the combination of local knowledge with scientific knowledge in dealing with climate change.

**Keywords** Climate change · Local knowledge and adaptation · Neglected and underutilised species

## 4.1 Climate Change and Agriculture

Climate change is projected to have a negative impact on the four pillars of food security: availability, access, utilisation and stability, and their interactions (IPCC, 2019; FAO et al., 2018). In a series of reports the Intergovernmental Panel on Climate Change (IPCC) and the FAO have considered the impact of climate change on agricultural yields (IPCC, 2015; FAO, 2016; FAO, 2018)) as well as the impact of agriculture on climate change as a contributor of greenhouse gases (FAO, 2015; IPCC, 2018; IPCC, 2019). Climate change in the first instance has a direct impact on agricultural yields through variations in water supply, in heat and in salinity and indirect effects through increasing insect infestations, plant diseases and reduction of pollinators (OECD, 2010; Almås & Campbell, 2012; Mendelsohn & Dinar, 2012; Maharjan & Joshi, 2013; Zolin & de Rodrigues, 2015; Das, 2016). High temperatures have been reported to cause physiological disorders in crops (Lee & Kader,

M. Blakeney (✉)
School of Law, The University of Western Australia, Crawley, WA, Australia
e-mail: michael.blakeney@uwa.edu.au



M. Blakeney, K. H. M. Siddique (eds.), *Local Knowledge, Intellectual Property and Agricultural Innovation*, https://doi.org/10.1007/978-981-15-4611-2_4

2000), in some cases temperature extremes are reported to predispose to aflatoxin contamination rendering food unsafe and requiring it to be discarded (Dorner, Cole, Sanders, & Blankenship, 1989). Delays in harvesting can cause the loss of crops through bird, rodent or insect attack and where the delay coincides with the onset of the rainy season, losses can be caused by rotting, the development of moulds and aflatoxin contamination, a major cause of food losses in cereals (Wareing, 2002; Lewis et al., 2005; Alakonya, Monda, & Ajanga, 2008).

Changes to climate will affect the seasons during which crops can be planted and harvested, as well as the varieties which can be cultivated. Extreme events, such as floods, storm surges, droughts and temperature spikes are increasingly affecting agricultural productivity, such as the late onset of monsoonal rains and extensive flooding in Kerala in 2018 and 2019 (Hindustan Times, 2018) and the strong El Niño effects in the Sahel region in 2015–2016 which resulted in crop failures in Ethiopia (FAO, 2016).

The IPCC (2019) has identified a number of studies of adverse agricultural impacts across the globe from climate change. Crop yield studies concerning India report a reduction in reduced wheat yields by 5.2% from 1981 to 2009 (Gupta, Somanathan, & Dey, 2017). Similar negative effects on wheat yields were noted in Australia (Innes et al., 2015). Studies of agriculture in the Hindu-Kush Himalayan region of India, Nepal, Pakistan, and China report disrupted agricultural yields from more frequent floods as well as prolonged droughts (Manzoor et al., 2013; Hussain et al., 2016). Similar results have been obtained in a number of South American studies (eg Saxena et al., 2016). Ketiem (2017) reported declines in yields of staple crops such as maize, wheat and sorghum across Africa.

The IPCC Special Report on Global Warming of 1.5 °C found that climate-related risks to food security are projected to increase with global warming (IPCC, 2018). The exclusive dependence of smallholder farmers on agriculture and their marginal location, and lack of essential farming resources makes them particularly vulnerable to climate change (Dasgupta, Mitra, & Sorger, 2018; Morton, 2007; Sietz, Choque, & Lüdeke, 2012).

Failure to maintain a low temperature of produce immediately after harvest is a major contributor to spoilage, particularly of perishable foods (World Bank, 2010). The heat and humidity associated with climate change also has an adverse effect on harvested crops during transport and storage by facilitating the growth of pest infestations, moulds and diseases (Moses, Jayas, & Alagusundaram, 2015).

## 4.2 Adaptation

The foundation international convention on climate change was the United Nations Framework Convention on Climate Change (FCCC), was negotiated at the United Nations Conference on Environment and Development (the Earth Summit) held in Rio de Janeiro from 3 to 14 June 1992. This convention entered into force on 21 March 1994 with the objective of stabilizing greenhouse gas concentrations in the atmosphere that would prevent dangerous anthropogenic interference with the climate system. To this end the parties committed themselves to limiting greenhouse gas emissions. This mitigation strategy was reaffirmed in the 1997 Kyoto Protocol and the 2015 Paris Agreement.

A complementary strategy, which was mentioned in the FCCC was adaptation to climate change through developing climate resilient agricultural practices. The IPCC in its 2019 report categorised adaptation measures as autonomous, incremental, and transformational (IPCC, 2019, 5–45). Autonomous adaptation is a spontaneous response to climate change, such as the variation to the timing of sowing and harvesting, the selection of short duration varieties, inter-cropping and changing cropping patterns. The experience of farmers making these changes, can make a valuable contribution to knowledge surrounding adaptation measures. Incremental adaptation focuses on improvements to existing resources and management practices, which will also be informed by local agricultural knowledge. Transformational adaptation involves a conscious response of agricultural practices in anticipation of or in response to climate change. The IPCC (2019, 5–45) refers, by way of example, to the shift from rice to sugar cane production among smallholder rice farmers in Northwest Costa Rica led to a due to decreasing market access and water scarcity (see Warner et al., 2015).

The earlier reports of the IPCC tended to focus upon mitigation and it was not until the Fifth IPCC Report in 2014 that adaptation strategies were considered in any detail (Noble et al., 2014). The IPCC Synthesis Report, 2014 noted that "indigenous, local and traditional knowledge systems and practices, including indigenous peoples' holistic view of community and environment, are a major resource for adapting to climate change, but these have not been used consistently in existing adaptation efforts." (IPCC, 2014). It has been pointed out that incorporating indigenous knowledge into climate change policies can lead to the development of effective mitigation and adaptation strategies that are cost-effective, participatory, and.

sustainable (Robinson & Herbert, 2001).

## 4.3 Local Knowledge and Climate Change Adaptation

As one of the principal impacts of climate change is upon rain-fed agriculture indigenous and traditional communities dependent on agricultural livelihoods are likely to be the most affected (Nyong, Adesina, & Elasha, 2007; Chang'a, Yanda, & Ngana, 2010; Sewando, Mutabazi, & Mdoe, 2016). Responding to climate stress local and indigenous peoples have developed coping mechanisms through experience and experimentation which has been transmitted both orally and in practice from one generation to the next. Their knowledge and practices would seem to be an obvious source of information for dealing with the contemporary challenge of climate change, particularly because they have to survive in marginal agricultural conditions (Bridges & McClatchey, 2009; Below, Schmid, & Sieber, 2014; Maldonado et al., 2016; Etchart, 2017; ILO, 2017).

Many rural farming communities have limited access to agricultural science information, a lack of skills to make use of the little information at their disposal (Pawluk, Sandor, & Tabor, 1992; Nyong et al., 2007). This may be due to high levels of illiteracy, lack of resources and inadequate support services (Dutta, 2009; Biggs et al., 2013). Thus, farmers tend to rely upon local and indigenous knowledge to make agricultural decisions concerning climate risks (Mapfumo, Mtambanengwe, & Chikowo, 2016).

However, this unique knowledge and experience concerning both adaptation and mitigation although valuable, has tended to be under-utilised (see Whitfield et al., 2015; Belfer, Ford, & Maillet, 2017), even by the IPCC in its early reports (Makondo & Thomas, 2018). The first serious consideration of the contribution which local or indigenous knowledge might have made to the development of mitigation or adaptation to climate change was in the IPCC's its Fourth Report (Parry et al., 2007). However, prior to this there had been a number of anthropological studies which demonstrated the usefulness of local ecological knowledge adapted to agricultural practices in agricultural crisis situations. Gadgil, Hemam, and Reddy (1998) refer to the maintenance of buffer areas of Sahelian range-lands, which are protected from grazing except in the case of emergencies. Berkes, Colding, & Folke, 2000 refer to limitations on the hunting of wading and seed dispersing birds in India in their breeding seasons.

Nyong et al. (2007) point out that farming communities have over the years relied on indigenous knowledge systems as a means of adapting to constantly climate variations. In a 2008 Issues Paper for the International Union for the Conservation of Nature (IUCN) it was pointed out that indigenous peoples of the tropical forest belt have developed specific coping strategies to deal with extreme variations of weather (Macchi et al., 2008). The Issues Paper identified as adaptation strategies:

(a) Crop diversification in order to minimize the risk of harvest failure;
(b) Changes of living area and a variety of movement patterns are used to deal with climatic variability.
(c) Change of hunting and gathering periods to adapt to changing animal migration and fruiting periods.
(d) Change of varieties and species to take account of new disease challenges.
(e) Changes in food storage methods, such as drying or smoking foods according to climate variability and corresponding availability of food.
(f) Changes in food habits, for example when the crops or cultivated plants are not producing good harvests, people will revert to gathering food in the forests. Or people who are close to a town might trade or barter with neighboring villages or traders/markets; and
(g) Forests as source of famine food in case of emergency (Macchi et al., 2008 at 40–41).

Surveys of farmers in Burkina Faso and Nigeria showed how indigenous methods of weather forecasting complemented the planning of agriculture in those countries (Roncoli, Ingram, & Kirshen, 2001; Ajibade & Shokemi, 2003).

In any event the knowledge of indigenous and local communities to manage climate change is sufficiently important to justify legislative intervention to preserve that knowledge (Hammer, Jintiach, & Tsakimp, 2013).

## 4.4 Governance

IPCC (2019, 5–111) pointed out that autonomous adaptation combined with indigenous and local knowledge were important for agricultural adaptation which necessitates farmer participation in governance structures, research, and the design of systems for the generation and dissemination of knowledge and technology, so that farmers needs and knowledge can be taken into consideration (IPCC, 2019, 5–111, citing Klenk et al., 2017). It instanced, by way of example, an innovation in terrace agriculture adopted by peasant farmers in Latin America which was developed on the basis of a local coping mechanism (see Bocco & Napoletano, 2017).

IPCC (2019, 5–57) maintains that the successful development of food systems under climate change conditions requires a multi-level institutional partnership that involves the cooperation of institutions and governance entities with local farmers. It instances this collaboration in the Nepal Local Adaptation Plan of Action, which promotes the combination of technological innovation with the tacit knowledge of farmers (citing Chhetri, Chaudhary, Tiwari, & Yadaw, 2012).

Community-based adaptation (CBA) is bottom-up approach to strengthen the climate adaptation capacity of local people (Ayers & Forsyth, 2009). Bocco and Napoletano (2017) refer to the modification of terrace agriculture to deal with climate change adopted by peasant farmers in Latin America as a locally-developed adaptation strategy.

Whether the harmonization of local knowledge with modern scientific innovations is feasible at a time when traditional and indigenous communities are being threatened by expropriation and colonization and deprivation of political rights (Thomas & Twyman, 2005; Ford et al., 2016) is an important question. One of the explanations for the reluctance of the IPCC to accept the importance of the contribution of local and traditional knowledge to climate change mitigation and adaptation may be the assertion that the recognition of indigenous rights, governance systems and laws is central to an understanding of their contribution (Thornton & Manasfi, 2010; Magni, 2017; Pearce, 2018). The involvement of indigenous communities in decision making is urged by Smith and Sharp (2012) as a matter of ethics and the preservation of those communities that are most vulnerable to climate change.

However, it is urged that the need for including indigenous and local peoples in climate mitigation policies and decision-making processes, there are procedural, conceptual and structural challenges (Brugnach, Craps, & Dewulf, 2017). The invariable inequality in the distribution of power sources between national governments on the one hand, and local indigenous communities on the other have tended to result in the exclusion of the latter (Roosvall & Tegelberg, 2013). Compounded by the lingering impacts of colonialism (Reo & Parker, 2013), these communities are often the passive recipients of international recipes for mitigation with little possibilities to defend their own legitimate interests and to contribute to climate change solutions. Brugnach et al. (2017) conclude that the main structural challenge resides in how to support the empowerment of indigenous communities to meaningfully influence policy development and decision-making processes for climate change mitigation. A similar challenge exists in relation to recognizing and valorizing the traditional knowledge of indigenous peoples (see Blakeney, 2016).

## 4.5 Seed Conservation

The Fifth IPCC Report noted that African women were particularly known to possess traditional knowledge to help maintain household food security, in times of drought and famine by relying upon indigenous plants that are more tolerant to droughts and pests (Parry et al., 2007, 866, referring to Ramphele (2004) and Eriksen (2005). By way of example, it instanced women in southern Sudan who are responsible for the selection of sorghum seeds saved for planting each year and who conserve a selection of seed varieties to ensure resistance to variable weather conditions (Parry et al., 2007, 866).

A number of studies have identified that locally developed seeds can often be more climate resilient than generic commercial varieties and are valuable in the protection of local agrobiodiversity, as well as providing a reservoir of useful climate resistant germplasm (Pautasso et al., 2013; Vasconcelos et al., 2013; van Niekerk & Wynberg, 2017).

It should be acknowledged that the germplasm flowing through exchange networks may come from a range of sources, including farmers' own fields and gardens and those of other farmers, local or district markets, NGOs and foundations, National Agricultural Research Systems, International Agricultural Research Centres, as well as commercial seed suppliers (Coomes et al., 2015). Furthermore, the circulation of farmer-saved seed is often embedded in social relations and institutions that constitute the social, economic and political fabric of rural life (see Sperling & McGuire, 2010; Thomas et al., 2011). Particularly, this last feature explains the characterization of farmer exchange systems as informal and lacking the reliability of the commercial seed sector. This has resulted in legislative intervention in some countries to ban such exchanges (see Wattnem, 2016). However, with the increasing realization of the utility of locally-saved seed at a time of climate change, much greater attention is being paid to this source of useful germplasm.

## 4.6 Agricultural Methods

IPCC (2019, 5–52) suggests that there are many current agricultural management practices that can be optimised and scaled up to advance adaptation. In a study of Tanzanian farming practices Scheba (2017) noted the role of local traditional knowledge of crop rotation, minimum tillage, mulching and the cultivation of cover crop in responding to climate stress.

Some of the traditional agricultural methods which had been criticized in the past are now being re-evaluated. For example, swidden agriculture had been dismissed as environmentally inappropriate "slash and burn". However, it is now acknowledged that in using ash as soil conditioner on degraded farmlands, swidden agriculture may have prevented desertification, and preserving forest habitat from uncontrolled wild-fires (Ando & Shinjo, 2017). Similarly, benign effects at a time of climate change has been found with fire management by Australian Aboriginal peoples (Bowman, Walsh, & Prior, 2004; Gott, 2005).

### *4.6.1 Cultivation of Neglected and Underutilized Species (NUS)*

Commercial seed is only a very small proportion of what farmers sow in developing countries (between 10 and 20%) (see Louwaars, de Boef, & Edme, 2013) and farmer networks are particularly important for the transmission of non-core crops. Local knowledge is playing an important role in identifying climate resilient plants that can constitute or supplement the food supply, particularly in marginal areas. (Baldermann et al., 2016; van der Merwe, Cloete, and van der Hoeven (2016).

The IPCC (2019) report referred to Adhikari, Hussain, and Rasul (2017) who noted the preference of farmers in the Rasuwa district, in the mid-hills of Nepal, prefer for local bean, barley, millet and maize, rather than commodity crops because they are more tolerant to water stress and extremely cold conditions.

In a study for the FAO, Li and Siddique (2018) sought to identify promising neglected and underutilized species (NUS) sometimes called "orphan crops" which are nutrition-dense and climate-resilient, as well as economically viable and locally available or adaptable as "Future Smart Food" (FSF) in the Asia-Pacific region. It was noted that studies had predicted that in South Asia climate change would result in a 14% decline in rice production, a 44–49% decline in wheat production, and a 9–19% fall in maize production (Nelson et al., 2009). They called for the collation of local and indigenous knowledge concerning useful NUS.

## 4.7 Cultural Factors

It is acknowledged that the cultivation and consumption of food crops is an intrinsically cultural process in which cultural identity and heritage plays an important role (Fuller & Qingwen, 2013). This may suggest that local and traditional communities lack the flexibility in shifting to more sustainable crops (eg Nielsen & Reenberg, 2010), but there is evidence from Africa that agricultural stresses on diets based on bananas or yams have them to shift to maize, which is more climate resilient, as well as to drought resistant cassava and millet, which supplements low maize yields (Chimhowu & Hulme, 2006; Rufino et al., 2013). Adger et al. (2013) suggest that local communities interpret and construct climate change trends and local indicators within a cultural setting and that mitigation and climate change adaptation should reflect modes of production, consumption and social organization.

Another cultural factor which is seen to be important for climate change mitigation and adaptation is the involvement of local and indigenous communities in the development of relevant strategies. A study of farmers in Zimbabwe observed that farmers were much more willing to accept the use of seasonal weather forecasts when they were associated with the forecasts of the traditional peoples (Patt & Gwata, 2002).

## 4.8 Local Knowledge and Scientific Knowledge

The successful adaptation to climate change in local and traditional communities requires an understanding of the processes of biophysical change and their interactions within socio-ecological systems. Thus, scientific data can be supplemented by local knowledge of practical adaptive practices to promote resilience to environmental changes (Murphy et al., 2016; Pearce, 2018; Kupika et al., 2019). For example, it was observed that in the Lushoto district, northern Tanzania, weather forecasts using indigenous knowledge were considered more reliable and specific to their location compared to scientific forecasts (Mahoo et al., 2015). To improve accuracy, the systematic documentation of that knowledge and the establishment of a framework for integrating it with the official weather forecasting authority was recommended and the need to establish an information dissemination network and entrench weather forecasting within the local government agricultural development programmes was recommended (Mahoo, 2013).

On the other hand, it should be recognized that the largely biological indicators on which local and indigenous knowledge is traditionally based are also subject to increased climate variability, limiting the scope of this knowledge as a basis for policy formulation and decision making (Mapfumo et al., 2016).

A number of studies urge the updating of the apparently outmoded and ineffective agricultural practices of traditional communities, by the application of modern agricultural techniques. However, it should also be acknowledged that local and traditional communities have useful knowledge to contribute to modern farming. Fitzgerald (1993) described maize farmers in Iowa as having become "deskilled" through passive reliance on seed firms for their hybrid seeds. Deskilling thus refers to the reliance of farmers on outside generated technologies and knowledge, losing their capacity to experiment with local and possibly more sustainable varieties (Scheinman, 2018). The introduction of genetically modified seeds with increasingly subtle new traits have compromised farmers' experience and reduced knowledge of seed management. Thus, for example, in India an estimated 30,000 traditional varieties of rice have been largely replaced by ten modern varieties that supply 75% of the country's rice crop and in Sri Lanka five modern varieties account for all of the rice grown in the country (Raeburn, 1995, 103). This genetic erosion was described by US Vice President Al Gore as the single most serious threat to the global food system (Gore, 1992).

Conserved local and traditional knowledge can act as a corrective to this agricultural deskilling (see Stone, 2007). Local knowledge may also explain agricultural practices which have negative impacts on local livelihoods and sustainable development (Nakashima et al., 2012).

Of course, any attempt to join scientific with local and indigenous knowledge systems may reflect the history of power relationships mentioned above between local and indigenous groups and the scientific establishment and that traditional knowledge is not uniform, even within small communities (Alexander et al., 2011).

## References

Adger, W. N., et al. (2013). Cultural dimensions of climate change impacts and adaptation. *Nature Climate Change, 3*(2), 112–117.

Adhikari, L., Hussain, A., & Rasul, G. (2017). Tapping the potential of neglected and underutilized food crops for sustainable nutrition security in the mountains of Pakistan and Nepal. *Sustainable, 9*(2), 291.

Ajibade, L. T., & Shokemi, O. (2003). Indigenous approaches to weather forecasting in Asa L.G.A. Kwara State, Nigeria. *Indilinga: African Journal of Indigenous Knowledge Systems, 2*(1), 37–44.

Alakonya, A. E., Monda, E. O., & Ajanga, S. (2008). Effect of delayed harvesting on maize ear rot in Western Kenya. *American-Eurasian Journal of Agricultural and Environmental, 4*(3), 372–380.

Alexander, C., et al. (2011). Linking indigenous and scientific knowledge of climate change. *Bioscience, 61*(6), 477–484.

Almås, R., & Campbell, H. (2012). *Rethinking agricultural policy regimes food security, climate change and the future resilience of global agriculture*. Bradford, England: Emerald Group Publishing.

Ando, K., & Shinjo, H. (2017). Slash-and-burn agriculture in Zambia. In S. Funakawa (Ed.), *Soils, ecosystem processes, and agricultural development* (pp. 253–274). Tokyo: Springer.

Ayers, J., & Forsyth, T. (2009). Community-based adaptation to climate change. *Environment: Science and Policy for Sustainable Development, 51*(4), 22–31.

Baldermann, S., et al. (2016). Are neglected plants the food for the future? *Critical Reviews in Plant Sciences, 35*(2), 106–119.

Belfer, E., Ford, J. D., & Maillet, M. (2017). Representation of indigenous peoples in climate change reporting. *Climatic Change, 145*, 57–70.

Below, T. B., Schmid, J. C., & Sieber, S. (2014). Farmers' knowledge and perception of climatic risks and options for climate change adaptation: A case study from two Tanzanian villages. *Regional Environmental Change, 15*(7), 1169–1180.

Berkes, F., Colding, J., & Folke, C. (2000). Rediscovery of traditional ecological knowledge as adaptive management. *Ecological Applications, 10*(5), 1251–1262.

Biggs, E. M., et al. (2013). Agricultural adaptation to climate change: Observations from the Mid-Hills of Nepal. *Climate and Development, 5*(2), 165–173.

Blakeney, M. (2016). The negotiations in WIPO for international conventions on traditional knowledge and traditional cultural expressions. In J. C. Lai & A. M. Dominicé (Eds.), *Intellectual property and access to Im/material goods* (pp. 227–256). Cheltenham, England: Edward Elgar.

Bocco, G., & Napoletano, B. M. (2017). The prospects of terrace agriculture as an adaptation to climate change in Latin America. *Geography Compass, 11*(10), 1–13.

Bowman, D. M. J. S., Walsh, A., & Prior, L. D. (2004). Landscape analysis of aboriginal fire management in Central Arnhem Land, North Australia. *Journal of Biogeography, 31*(2), 207–223.

Bridges, K. W., & McClatchey, W. C. (2009). Living on the margin: Ethnoecological insights from Marshall islanders at Rongelap Atoll. *Global Environmental Change, 19*(2), 140–146.

Brugnach, M., Craps, M., & Dewulf, A. (2017). Including indigenous peoples in climate change mitigation: Addressing issues of scale, knowledge and power. *Climatic Change, 140*, 19–32.

Chang'a, L. B., Yanda, P. Z., & Ngana, J. (2010). Indigenous knowledge in seasonal rainfall prediction in Tanzania : A case of the South-Western Highland of Tanzania. *Journal of Geography and Regional Planning, 3*(4), 66–72.

Chhetri, N., Chaudhary, P., Tiwari, P. R., & Yadaw, R. B. (2012). Institutional and technological innovation. Understanding agricultural adaptation to climate change in Nepal. *Applied Geography, 33*(1), 142–150.

Chimhowu, A., & Hulme, D. (2006). Livelihood dynamics in planned and spontaneous resettlement in Zimbabwe: Converging and vulnerable. *World Development, 34*(4), 728–750.

Coomes, O. T., et al. (2015). Farmer seed networks make a limited contribution to agriculture? Four common misconceptions. *Food Policy, 56*, 41–50.

Das, H. P. (2016). *Climate change and agriculture: Implication for global food security*. London: CRC Press.

Dasgupta, P., Mitra, T., & Sorger, G. (2018). Harvesting the Commons. *Environmental and Resource Economics, 72*(3), 613–636.

Dorner, J. W., Cole, R. J., Sanders, T. H., & Blankenship, P. D. (1989). Interrelationship of kernel water activity, soil temperature, maturity, and phytoalexin production in preharvest aflatoxin contamination of drought-stressed peanuts. *Mycopathologia, 105*, 117–128.

Dutta, R. (2009). Information needs and information-seeking behaviour in developing countries: A review of the research. *The International Information and Library Review, 41*(1), 44–51.

Eriksen, S. (2005). The role of indigenous plants in household adaptation to climate change: The Kenyan experience. In P. S. Low (Ed.), *Climate change and Africa* (pp. 248–259). Cambridge, UK: CUP.

Etchart, L. (2017). The role of indigenous peoples in combating climate change. *Palgrave Communications, 3*, 17085.

FAO. (2015). *Food wastage footprint and climate change*. Rome. Available at http://www.fao.org/fileadmin/templates/nr/sustainability_pathways/docs/FWF_and_climate_change.pdf. Accessed 24 May 2018.

FAO. (2016). *2015–2016 El Niño – Early action and response for agriculture, food security and nutrition – UPDATE #10*. Available at http://www.fao.org/3/a-i6049e.pdf. Accessed 24 Oct 2019.

FAO. (2018). *The future of food and agriculture: Trends and challenges*. Available at http://www.fao.org/3/a-i6583e.pdf. Accessed 24 Oct 2019.

FAO, IFAD, UNICEF, WFP, & WHO. (2018). *The state of food security and nutrition in the world 2018, building climate resilience for food security and nutrition*. Available at www.fao.org/publications. Accessed 24 October 2019.

Fitzgerald, D. (1993). Farmers deskilled: Hybrid corn and Farmers' work. *Technology and Culture, 34*(2), 324–343.

Ford, J., et al. (2016). Adaptation and indigenous peoples in the United Nations framework convention on climate change. *Climatic Change, 139*(3–4), 429–443.

Fuller, T., & Qingwen, M. (2013). Understanding agricultural heritage sites as complex adaptive systems: The challenge of complexity. *Journal of Resources and Ecology, 4*(3), 195–201.

Gadgil, M., Hemam, N. S., & Reddy, B. M. (1998). People, refugia and resilience. In F. Berkes & C. Folke (Eds.), *Linking social and ecological systems: Management practices and social mechanisms for building resilience* (pp. 30–47). Cambridge, UK: CUP.

Gore, A. (1992). *Earth in the balance. Ecology and the human Spirit*. New York: Houghton Mifflin.

Gott, B. (2005). Aboriginal fire management in south eastern Australia: Aims and frequency. *Journal of Biogeography, 32*(7), 1203–1208.

Gupta, R., Somanathan, E., & Dey, S. (2017). Global warming and local air pollution have reduced wheat yields in India. *Climate Change, 140*, 593–604.

Hammer, C., Jintiach, J. C., & Tsakimp, R. (2013). Practical developments in law science and policy: Efforts to protect the traditional knowledge and practices of the Shuar, an indigenous people in the Ecuadorian Amazon. *Policy Sciences, 46*, 125–141.

Hindustan Times. (2018). August 27. Available at https://www.hindustantimes.com/india-news/kerala-floods-damage-45-000-ha-of-farm-fields-paddy-banana-spices-among-worst-hit/story-bF0Iully9ppN6vWnuDcGHL.html. Accessed 24 Oct 2019.

Hussain, A., et al. (2016). Household food security in the face of climate change in the Hindu-Kush Himalayan region. *Food Security, 8*(5), 921–937.

ILO. (2017). *Indigenous peoples and climate change: From victims to change agents through decent work* (p. 2017). Geneva, Switzerland: ILO.

Innes, P. J., et al. (2015). Effects of high-temperature episodes on wheat yields in New South Wales, Australia. *Agricultural and Forest Meteorology, 208*, 95–107.

IPCC. (2014). *Climate change 2014* (Synthesis Report). Geneva, Switzerland: WHO and Gland UNEP.

IPCC. (2018). *Global IPCC warming of 1.5 °C* (An IPCC special report on the impacts of global warming of 1.5°C above pre-industrial levels and related global greenhouse gas emission pathways, in the context of strengthening the global response to the threat of climate change, sustainable development, and efforts to eradicate poverty). Available at http://report.ipcc.ch/sr15/pdf/sr15_spm_final.pdf. Accessed 24 October 2019.

IPCC. (2019). *IPCC special report on climate change, desertification, land degradation, sustainable land management, food security, and greenhouse gas fluxes in terrestrial ecosystems*. Available at https://www.ipcc.ch/report/srccl/. Accessed 24 Oct 2019.

Ketiem, P. (2017). Integration of climate change information into drylands crop production practices for enhanced food security: A case study of lower Tana Basin in Kenya. *African Journal of Agricultural Research, 12*(2), 1763–1771.

Klenk, N., et al. (2017). Local knowledge in climate adaptation research: Moving knowledge frameworks from extraction to co-production. *Wiley Interdisciplinary Reviews: Climate Change, 8*(5), e475.

Kupika, O. L., et al. (2019). Local ecological knowledge on climate change and ecosystem-based adaptation strategies promote resilience in the Middle Zambezi biosphere reserve, Zimbabwe. *Hindawi Scientifica*. Article ID 3069254.

Lee, S. K., & Kader, A. A. (2000). Pre-harvest and postharvest factors influencing vitamin C content of horticultural crops. *Postharvest Biology and Technology, 20*(3), 207–220.

Lewis, L., et al. (2005). Aflatoxin contamination of commercial maize products during an outbreak of acute aflatoxicosis in Eastern and Central Kenya. *Environmental Health Perspectives, 113*(12), 1763–1767.

Li, X., & Siddique, H. M. K. (Eds.). (2018). *Future smart food. Rediscovering hidden treasures of neglected and underutilized species for Zero Hunger in Asia*. Bangkok, Thailand: FAO.

Louwaars, N., de Boef, W., & Edme, J. (2013). Integrated seed sector development in Africa: A basis for seed policy and law. *Journal of Crop Improvement, 27*, 186–214.

Macchi, M., et al. (2008). *Indigenous and traditional peoples and climate change, issues paper*. Gland, Switzerland: IUCN.

Magni, G. (2017). Indigenous knowledge and implications for the sustainable development agenda. *European Journal of Education, 52*(4), 437–447.

Maharjan, K. L., & Joshi, N. P. (2013). *Climate change, agriculture and rural livelihoods in developing countries*. Tokyo, Japan: Springer.

Mahoo, H., et al. (2015). Integrating indigenous knowledge with scientific seasonal forecasts for climate risk management in Lushoto district in Tanzania (CCAFS Working Paper no. 103, CGIAR Research Program on Climate Change, Agriculture and Food Security (CCAFS)). Copenhagen, Denmark.

Makondo, C. C., & Thomas, D. S. G. (2018). Climate change adaptation: Linking indigenous knowledge with western science for effective adaptation. *Environmental Science and Policy, 88*, 83–91.

Maldonado, J., et al. (2016). Engagement with indigenous peoples and honoring traditional knowledge systems. *Climatic Change, 135*(1), 111–126.

Manzoor, M., et al. (2013). Historical analysis of flood information and impacts assessment and associated response in Pakistan (1947–2011). *Research Journal of Environmental and Earth Sciences, 5*, 139–146.

Mapfumo, P., Mtambanengwe, F., & Chikowo, R. (2016). Building on indigenous knowledge to strengthen the capacity of smallholder farming communities to adapt to climate change and variability in southern Africa. *Climate and Development, 8*(1), 72–82.

Mendelsohn, R., & Dinar, A. (2012). *Handbook on climate change and agriculture*. Cheltenham, England: Edward Elgar Publishing.

Morton, J. F. (2007). The impact of climate change on smallholder and subsistence agriculture. *Proceedings of the National Academy of Sciences, 104*, 19680–19685.

Moses, J. A., Jayas, D. S., & Alagusundaram, K. (2015). Climate change and its implications on stored food grains. *Agricultural Research, 4*(1), 21–30.

Murphy, C., et al. (2016). Adapting to climate change in shifting landscapes of belief. *Climatic Change, 134*(1), 101–114.

Nakashima, D. J., et al. (2012). *Weathering uncertainty: Traditional knowledge for climate change assessment and adaptation*. Paris: UNESCO.

Nelson, G. C., et al. (2009). *Climate change: Impact on agriculture and costs of adaptation* (Food Policy Report). Washington, DC: International Food Policy Research Institute (IFPRI).

Nielsen, J. Ø., & Reenberg, A. (2010). Cultural barriers to climate change adaptation: A case study from northern Burkina Faso. *Global Environmental Change, 20*(1), 142–152.

Noble, I. R., et al. (2014). *Climate change 2014: Impacts, adaptation, and vulnerability* (Part A: Global and Sectoral Aspects. Contribution of Working Group II to the Fifth Assessment Report of the Intergovernmental Panel on Climate Change) (pp. 833–868). Cambridge, UK: CUP.

Nyong, A., Adesina, F., & Elasha, B. O. (2007). The value of indigenous knowledge in climate change mitigation and adaptation strategies in the African Sahel. *Mitigation and Adaptation Strategies for Global Change, 12*(5), 787–797.

OECD. (2010). *Climate change and agriculture impacts, adaptation and mitigation*. Paris: OECD.

Parry, M. L., et al. (2007). Cross-chapter case study. In *Climate change 2007: Impacts, adaptation and vulnerability* (Contribution of Working Group II to the Fourth Assessment Report of the Intergovernmental Panel on Climate Change) (pp. 843–868). Cambridge, UK: CUP.

Patt, A., & Gwata, C. (2002). Effective seasonal climate forecast applications: Examining constraints for subsistence farmers in Zimbabwe. *Global Environmental Change, 12*(3), 185–195.

Pautasso, M., et al. (2013). Seed exchange networks for agrobiodiversity conservation. A review. *Agronomy for Sustainable Development, 33*, 151–175.

Pawluk, R. R., Sandor, J. A., & Tabor, J. A. (1992). The role of indigenous soil knowledge in agricultural development. *Journal of Soil and Water Conservation, 47*(4), 298–302.

Pearce, T. C. L. (2018). Incorporating indigenous knowledge in research. In R. McLeman & F. Gemenne (Eds.), *Routledge handbook of environmental migration and displacement* (pp. 125–134). New York/London: Taylor & Francis Group.

Raeburn, P. (1995). *The last harvest: The genetic gamble that threatens to destroy American agriculture*. Lincoln, England: University of Nebraska Press.

Ramphele, M. (2004). Women's indigenous knowledge: Building bridges between the traditional and the modern. In World Bank (Ed.), *Indigenous knowledge: Local pathways to development* (pp. 13–17). Washington, DC: The World Bank.

Reo, N. J., & Parker, A. K. (2013). Re-thinking colonialism to prepare for the impacts of rapid environmental change. *Climatic Change, 120*(3), 671–682.

Robinson, J., & Herbert, D. (2001). Integrating climate change and sustainable development. *International Journal of Global Environmental Issues, 1*(2), 130–148.

Roncoli, C., Ingram, K., & Kirshen, P. (2001). The costs and risks of coping with drought: Livelihood impacts and farmers' responses in Burkina Faso. *Climate Research, 19*, 119–132.

Roosvall, A., & Tegelberg, M. (2013). Framing climate change and indigenous peoples: Intermediaries of urgency, spirituality and de-nationalization. *International Communication Gazette, 75*(4), 392–409.

Rufino, M. C., et al. (2013). Transitions in agro-pastoralist systems of East Africa: Impacts on food security and poverty. *Agriculture, Ecosystems and Environment, 179*, 215–230.

Saxena, A. K., et al. (2016). Indigenous food systems and climate change: Impacts of climatic shifts on the production and processing of native and traditional crops in the Bolivian Andes. *Frontiers in Public Health, 4*, 20.

Scheba, A. (2017). Conservation agriculture and sustainable development in Africa: Insights from Tanzania. *Natural Resources Forum, 41*(4), 209–219.

Scheinman, E. (2018). *Contested futures – Traditional rice as a keystone of alternative agriculture movements. Agricultural knowledge, seed systems and agrarian reform in South Karnataka, India*. MES thesis, Queen's University, Kingston, Ontario, Canada.

Sewando, P. T., Mutabazi, K. D., & Mdoe, N. Y. S. (2016). Vulnerability of agro-pastoral farmers to climate risks in northern and central Tanzania. *Development Studies Research, 3*(1), 11–24.

Sietz, D., Choque, S. E. M., & Lüdeke, M. K. B. (2012). Typical patterns of smallholder vulnerability to weather extremes with regard to food security in the Peruvian Altiplano. *Regional Environmental Change, 12*, 489–505.

Smith, H. A., & Sharp, K. (2012). Indigenous climate knowledges. *Wiley Interdisciplinary Reviews: Climate Change, 3*(5), 467–476.

Sperling, L., & McGuire, S. (2010). Understanding and strengthening informal seed markets. *Experimental Agriculture, 46*, 119–136.

Stone, G. D. (2007). The birth and death of traditional knowledge: Paradoxical effects of biotechnology in India. In C. McManis (Ed.), *Biodiversity and the law: Intellectual property, biotechnology and traditional knowledge* (pp. 207–238). London: Earthscan.

Thomas, D. S. G., & Twyman, C. (2005). Equity and justice in climate change adaptation amongst natural-resource-dependent societies. *Global Environmental Change, 15*(2), 115–124.

Thomas, M., et al. (2011). Seed exchanges, a key to analyze crop diversity dynamics in farmer-led on-farm conservation. *Genetic Resources and Crop Evolution, 58*, 321–338.

Thornton, T. F., & Manasfi, N. (2010). Adaptation – Genuine and spurious: Demystifying adaptation processes in relation to climate change. *Environment and Society, 1*(1), 132–155.

van der Merwe, J. D., Cloete, P. C., & van der Hoeven, M. (2016). Promoting food security through indigenous and traditional food crops. *Agroecology and Sustainable Food Systems, 40*, 830–847.

van Niekerk, J., & Wynberg, R. (2017). Traditional seed and exchange systems cement social relations and provide a safety net: A case study from KwaZulu-Natal, South Africa. *Agroecology and Sustainable Food Systems, 41*(9–10), 1–25.

Vasconcelos, A. C. F., et al. (2013). Landraces as an adaptation strategy to climate change for smallholders in Santa Catarina. *Southern Brazil, Land Use Policy, 34*, 250–254.

Wareing, P. (2002). Pests of durable crops – moulds. In P. Golob, G. Farrell, & J. E. Orchard (Eds.), *Crop post-harvest: Science and technology* (Principles and Practice) (Vol. 1, pp. 120–130). Oxford, UK: Blackwell Sciences.

Warner, B. P., et al. (2015). Limits to adaptation to interacting global change risks among smallholder rice farmers in Northwest Costa Rica. *Global Environmental Change, 30*, 101–112.

Wattnem, T. (2016). Seed laws, certification and standardization: outlawing informal seed systems in the Global South. *Journal of Peasant Studies, 43*, 850–867.

Whitfield, S., et al. (2015). Critical reflection on knowledge and narratives of conservation Agriculture. *Geoforum, 60*, 133–142.

Zolin, C. A., & de Rodrigues, A. R. R. (2015). *Impact of climate change on water resources in agriculture*. London: CRC Press.

# Chapter 5
# Plant Variety Rights and Food Security

**Michael Blakeney**

**Abstract** This chapter explores the origins of plant variety rights protection and the evolution of the UPOV Convention. It refers to the various revisions of UPOV and its relationship with the WTO TRIPS Agreement. It explores the impact on plant variety protection of the recent development in some patent laws which allow for the patenting of plant breeding methods. Plant variety protection in developing countries is discussed, as well as the environmental and ecological impacts of plant variety laws.

**Keywords** Plant variety rights protection · UPOV convention · UPOV 1991 · Patenting of plant breeding methods · Environment

## 5.1 Introduction

At the heart of agriculture is a complex interrelationship between farmers and seed producers. Farmers seek access to seed which is suitable for their environmental circumstances and which will secure them reasonable economic returns for their endeavours. Seed producers seek the maximum returns for their investment in seed breeding through continuing sales of their seed to farmers. The utilization of local knowledge of suitable land-races was originally the basis for selection which farmers made of seed which was appropriate for the climatic conditions and soil types within which they had to work and which produced crops that appealed to consumers. Future plantings were secured by the retention of harvested seed and by purchase or exchange of seed with other local farmers. With the application of Gregor Mendel's theories to agriculture in the early 1900s the establishment of plant breeding on genetic principles became feasible (see Allard, 1960, ch.1; Harwood, 2000; Palladino, 1994). Prior to this time farmers had, of course, selected and harvested seeds from plants which had desirable traits, such as disease resistance, and

M. Blakeney (✉)
School of Law, The University of Western Australia, Crawley, WA, Australia
e-mail: michael.blakeney@uwa.edu.au



M. Blakeney, K. H. M. Siddique (eds.), *Local Knowledge, Intellectual Property and Agricultural Innovation*, https://doi.org/10.1007/978-981-15-4611-2_5

suitability to their local conditions, without being aware of the genetic mechanisms which produced these results (see Murphy, 2007). The significance of Mendel's theories is that it made possible the establishment of a plant breeding industry, which shifted agricultural innovation from the farmer to corporations (Kloppenburg, 2004; Palladino, 2002; Silva Dias & Ryder, 2012, cf).

To secure continuing purchases from farmers, seed companies had to offer varieties which produced yields which were superior to those superior to those which had traditionally been cultivated and which through legal protection or technological manipulation obliged farmers to purchase further supplies of seed. The development of new varieties required the identification of useful germplasm, often which had been identified and conserved *in situ* by traditional farmers, assembling diverse genetic material, crossing the genetic material to generate variation, selecting recombinations from the variation, and stabilising the preferred characteristics as a new variety. The object of varietal development was to embody selected traits, such as resistance to disease or climate stresses or high yields which would appeal to farmers. Seed breeding typically involves a three-stage process that sequentially produces 'foundation seed', 'certified seed' and 'commercial seed', usually dependent upon the regulatory system which applies in the relevant country (Gregg et al., 1980; Jaffee and Strivastava, 1994). The seed breeder has to ensure that the seed meets the claims which are made for the variety and through preservation measures, such as drying, cleaning, and chemical treatment, the viability of the seed is maintained between production and eventual utilisation. Finally, the marketing and distribution of the seed requires promotion to farmers of the characteristics and agronomic performance of available varieties through advertising and field demonstrations and distribution to farmers to ensure that the seed of the variety reaches the farmer at the correct time.

The development of the seed industry has evolved from being largely based upon traditional farmers as the primary source for varietal improvement, followed by a middle stage dominated by public sector breeding with the current structure in which private sector companies are pre-eminent. The expense of varietal development is often cited for the withdrawal of public sector from the seed industry (Knudson, 1990; Jaffee and Strivastava, 1992). This expense is also reflected in the progressive market concentration in the seed industry and its focus upon focus on high-value commercial crops and hybrids. Although the commercial seed industry accounts for approximately one third of the global market by value, in developing countries commercial seed providers are relatively insignificant, for example in India only 7% of wheat seed and 13% of rice seed in India is sourced from the formal seed sector (comprising both public and private sector bodies) (Cromwell, Friis-Hansen, & Turner, 1992; Morris, 1998 and ten Kate and Laird (1999) estimate that in many parts of Africa and Asia over 80% of total farmers' seed requirement are met from outside the formal sector.

Paralleling the developing of a seed breeding industry has been the growth primarily of industrialised countries of large scale, mechanised agriculture in which seed saving and cleaning by farmers was apparently less convenient than the purchase of farm-ready seed from dealers (see Fowler, 1994). The involvement of large corporations in the plant breeding industry has inevitably resulted in a high degree of market concentration. Over the past three decades, a series of mergers and acquisitions created the "Big Six": Monsanto, Bayer, BASF, Syngenta, Dow and DuPont. After the introduction of herbicide tolerance genes, it made sense for firms to combine activities in pesticides, seed breeding and GM technology. The recent merger wave reduces the number of major firms to four (Bayer-Monsanto, DowDuPont/Corteva, ChemChina-Syngenta, BASF). Market concentration in the seed market has been estimated to exceed 60%, although the OECD (2018) has identified concentration levels at 100% in some markets (eg Bayer and Monsanto in the Mexican cotton market). It is uncertain whether these high concentration levels have reduced innovation in those concentrated markets (OECD, 2018) but the business model of these companies requires the sale of herbicide-tolerant seed to stimulate herbicide sales, which in turn would stimulate sales of herbicide-tolerant seed. Intellectual property protection is required to protect the investments in seed R & D.

An important question, particularly for developing countries, which are considered below, is whether the legislative protection of plant varieties has resulted in an increase in the number of varieties which are available to farmers. Even in industrialised countries, the evidence is equivocal. Lesser (1990) claims that "the availability of PBRs has increased the number of private sector breeders, as well as the number of varieties released and planted" (Lesser, 1990, 1991) whereas Butler and Marion (1985) Kloppenburg (2004) and Rangnekar (2001) consider the R&D impact of plant variety rights protection is modest at best. An intermediate position is that there is an uneven impact of plant variety protection depending upon the crop type (Perrin, Kunnings, & Ihnen, 1983; Ramaswami, 2000; Rangnekar, 2000). It has been pointed out that even in the USA, there is little evidence that plant variety rights protection has resulted in an increase in the range of plant materials available to farmers or to an increased rate of innovation (Alston & Venner, 2002; Carew & Devados, 2003; Kolady & Lesser, 2009). There is limited experience from developing countries on the impact of plant variety protection. Even if it can be shown that the protection of plant varieties has resulted in an increase in new varieties, it has been questioned whether this results in agronomically superior products or represents merely cosmetic changes in varieties or the institution of planned obsolescence (Rangnekar, 2002).

The range of crops focussed on by the large plant breeding companies do not cater for the wider needs of the farming populations in developing countries. These farmers are not a market with significant purchasing power and a substantial segment of the developing country market is supplied by non-commercial transactions, such as exchange of seed. This raises the question as to whether the absence of effective demand forestalls the supply of suitable plant varieties, or does the lack of suitable varieties pre-empt the demand for proprietary seed.

## 5.2 First Moves to Provide Intellectual Property Protection for Plants

The establishment of a plant breeding industry was potentially undermined by the ease with which new varieties could be replicated by competitors, since plants could easily be produced from the seeds or tubers of desirable varieties. The original breeders of these varieties sought protection for the investments which they had made in originating the varieties (Kevles, 2011). Intellectual property was called in aid to protect these investments.

The first legislative proposal for the protection of agricultural innovations was the Papal States Edict of 3 September 1833 concerning the declarations of ownership of new inventions and discoveries in the fields of the technological arts and agriculture, but this general measure was never implemented (Heitz, 1991). The first national proposal that foreshadowed the protection of agricultural innovations under patent law was the introduction, in the United States Congress of 1906, of a "Bill to amend the laws of patents in the interest of the originators of horticultural products." This bill was unsuccessful, as were similar bills introduced in 1907, 1908 and 1910. It was not until the Townsend-Parnell Act of 1930, the "Plant Patent Act," that agricultural innovations were recognised by Congress. This statute endures as sections 161–164 of the current United States patent law (35 U.S.C. §§ 161–164, 2000).

Section 161 of the Plant Patent Act confined protection to asexually reproduced plants, because of the view that sexually reproduced varieties lacked stability (see Williams, 1983). The section also excluded tuber-propagated plants principally because of a concern that this would lead to monopolies in basic foodstuffs, such as potatoes (Sanderson, 2017).

Applicants for plant patents were accordingly required to asexually reproduce the plant in relation to which protection was sought, in order to demonstrate the stability of the characteristics that were claimed.

Section 161 also required that eligible new varieties should be "distinct." The statute did not define this requirement, although the Senate Committee Report accompanying the Act stated that "in order for a new variety to be distinct it must have characteristics clearly distinguishable from those of existing varieties" and that it was not necessary for the new variety to constitute "a variety of a new species" (quoted in Rossman, 1935).

Legislation similar to the U.S. Plant Patents Act was adopted in Cuba, 1937; South Africa, 1952 and the Republic of Korea, 1973, in an endeavour by those countries to align their patent systems with that of the United States (Heitz, 1991, 23).

In other parts of the world patent protection was not considered apposite for the protection of new plant varieties. A new variety was generally characterised as a discovery, rather than as an invention (MacLeod, 1996) and the role of plant breeders in selecting desirable traits was considered to be obvious and not involving the "inventive step" of patent law (Pottage & Sherman, 2007).

## 5.3 The Road to UPOV

The first international instrument for *sui generis*, or specially created protection for plant varieties was the International Convention for the Protection of New Varieties of Plants (UPOV Convention) which was signed in Paris in 1961.

In Europe, the first steps toward UPOV were a French decree of 5 December 1922 which introduced a Register for Newly-bred Plants and a similar system of seed certification was established by the Netherlands in 1932. The first national statute that clearly anticipated the UPOV Convention was the Czech Law of 1921 on the Originality of Types, Seeds and Seedlings and the Testing of Horticultural Types. It provided that registration of plant seed types entitled the registrant to place its material in commerce under a registered indication. The horticulturalist or producer who produced the original material obtained the exclusive right to make use of a registered trade mark covering the type.

The U.S. Plant Patents Act 1930 was emulated in the draft Seeds and Seedlings Law, which was submitted to the German Parliament in 1930 (*GRUR* 244 [1930]), the year in which the US Act was adopted. The German legislation provided protection to plant breeders for new varieties that were distinguishable from existing varieties in characteristics that were inheritable or transferable by vegetative propagation. The German Law denied protection to a variety obtained by a mere selection without important or substantial improvement of an existing protected variety. The Law also authorised the registration of protected varieties as trademarks. However, this draft Law was never adopted by the German Parliament.

A more obvious precursor to the UPOV Convention was the German Law of 27 June 1953, on the Protection of Varieties and the Seeds of Cultivated Plants. Article 1 of this statute stated that the purpose of protection was to promote the creation of useful (*wetvoll*) new varieties of cultivated plants. An exception was provided for non-food plants and varieties intended for export. A precondition for protection was that a variety should be "individualised" and stable. This anticipated the UPOV requirements of distinctiveness and stability. The registered owner of a protected variety had the exclusive right to produce and sell seed of the variety. The Law also permitted the use of a protected variety for the creation of new varieties. Also anticipating UPOV was the requirement that anyone who marketed seed of the protected variety was obliged to use the registered designation for the variety. As with UPOV, where under the German Law the variety designation was a registered trade mark, the trade mark proprietor could not object to the use of the designation where such use was compulsory.

As with other categories of intellectual property, a key role in the inclusion of agricultural innovations within the international regulatory regime was played by industry associations. The Congrès pomologique de France, held in 1911, had called for special protection of plant varieties. The International Union of the Horticultural Profession, also considered the matter at its Congresses in Luxemburg (1911), London (1912) and Ghent (1913). The International Institute of Agriculture in its 1927 Congress had stated that the protection of a denomination was insufficient and that a way had to be found to require "any grower who engaged in reproduction of those breeds for the purposes of sale to pay a royalty to the producer"(quoted in UPOV, 1987, 80).

The International Federation of Breeders of Staple Crops had, in its 1931 conference, expressed the hope that the legal status of new varieties should be assimilated to that of industrial inventions. Discussions concerning the creation of a new organization to agitate for the promulgation of an international legal regime for the protection of plant varieties occurred at the meetings of the International Breeders' Congress at Leeuwarden in 1936 and the 1937 Conference of the International Organization of Agricultural Industries, also held in the Netherlands. The direct result of these discussions was the foundation in Amsterdam, on November 17, 1938, of the International Association of Plant Breeders for the Protection of Plant Varieties (ASSINSEL). The first ASSINSEL Congress, held in Paris on 8–9 July 1939 adopted a three-point resolution:

To accept internationally the filing of trademarks and appellations as a means of protection (pending introduction of a patent);
To adopt the principle of a licence, to be drawn up by ASSINSEL for the purposes of multiplication and sale; and.
To accept internationally the definition of the word 'original' [as] seed produced, offered or sold by the breeder of the variety or under his control by his licensees or successors in title.

The Second World War interrupted these developments. At its Semmering Congress in June, 1956, a resolution of ASSINSEL called for an international conference to promulgate an international system for the protection of plant varieties. The French Government had been approached by ASSINSEL, because it had indicated a favourable attitude. Invitations were issued to 12 Western European countries[1] to attend a diplomatic conference in Paris, from 7 to 11 May 1957. The notes of invitation to the conference referred to the conclusions that had been reached at the 1954 conference on the Development of Seed Production and Trade, held in Stockholm, that there should be an international agreement favourable to the protection of new plant varieties.

## 5.4 The Paris Conferences on Special Protection of 1957 and 1961

The 1957 diplomatic conference in Paris was to consider establishing an international regime for the protection of plant varieties. Participation was limited by the French to those states who were known to share its own concerns on this subject.

[1] *I.e.* Austria, Belgium, Denmark, Finland, Federal Republic of Germany, Italy, the Netherlands, Norway, Spain, Sweden, Switzerland and the UK.

Thus, the United States was not invited because it had "confined itself to plant patents for vegetatively reproduced varieties, with at best only a minor part to play as foods" (UPOV, 1987, 82).

The conclusions of the 1957 Conference were set out in its Final Act, adopted on 11 May 1957. This instrument recognised the legitimacy of breeders' rights and established, as the preconditions for protection, that a variety had to be distinct from pre-existing varieties and sufficiently homogenous and stable in its essential characteristics. It defined the rights of the breeder and acknowledged the principle of the independence of protection in each country. It proposed that these principles be enshrined in an international Convention and that a Drafting Committee and a Committee of Experts be established.

Following three meetings of the Drafting Committee and two meetings of Committees of Experts, the second session of the Conference was held in Paris from 21 November to 2 December, 1961. An International Convention for the Protection of New Varieties of Plants (UPOV) was presented for the Consideration of the Conference. An important question debated there was whether the UPOV Convention would be compatible with the Paris Convention on the Protection of Industrial Property. The debate on that subject produced the inclusion of Article 2(1), which stated that "each Member of the [UPOV] Union may recognise the right of the breeder…by the grant of a special title of protection or a patent. Nevertheless, a Member State of the Union, whose national law admits of protection under both these forms may only provide one of them for one and the same genus or species."

Article 4(1) applied the draft UPOV Convention to "all botanical genera and species," but it was envisaged that the Convention would have a gradual introduction. A list of 13 genera was annexed to the Convention: wheat, barley, oats or rice, maize, potato, peas, beans, Lucerne, red clover, ryegrass, lettuce, apples, roses or carnations. Article 4(3) required each member State on entry into force of the Convention to apply it to at least five genera from this list and, within 8 years, to all the listed genera.

The UPOV Convention was signed on 2 December 1961 by the representatives of Belgium, France, the Federal Republic of Germany, Italy and the Netherlands. On 26 November 1962, the signatures of Denmark and the United Kingdom were added, followed by Switzerland on 30 November 1962. The Convention entered into force on 10 August 1968, following its ratification by Netherlands, the Federal Republic of Germany and the United Kingdom. Denmark deposited its instrument of ratification on 6 September 1968 and France on 3 September 1971. Sweden deposited an instrument of accession on 17 November 1971. Thus UPOV was from its inception, seen to be a proxy for the European seed breeding industry and its subsequent evolution involved stuttering attempts to broaden the membership of the convention.

The first step, by the revision of 1972 was to modify UPOV's budget by creating classes of membership envisaging lower subscriptions to be paid by developing countries and LDCs.

## 5.5 Revision of 1978

In an endeavour to broaden the membership of the Convention, invitations were widely circulated, to permit non-member states to participate as observers. In the end, some 27 non-member states attended, including the U.S. and a number of developing countries. One result was an amendment of Article 2 of the Convention to permit the accession of countries like the United States, which had laws allowing the double protection of varieties under patent and *sui generis* laws (UPOV, 1978; Byrne, 1991, 13).

The list of genera, annexed to the 1961 Convention was removed. This list had contained mainly species from temperate climates. Under the new Article 4, member states agreed to apply the Convention to at least five genera or species, rising to 24 genera or species within 8 years. Additionally, a grace period was introduced to permit the marketing of varieties 12 months prior to an application for plant variety protection being made. The revised Convention attracted the ratification of the United States on 12 November 1980, although by 1 January 1990 there were still only 18 members all of which were developed countries, mainly from Europe.

## 5.6 Revision of 1991

A further broadening of the UPOV Convention occurred with the 1991 Revision. The 1991 Act requires states to protect at least 15 plant genera or species upon becoming members of the Act, and to extend protection to all plant varieties within 10 years. In response to demands from breeders in industrialized counties, the 1991 Act required signatory states to make dual protection mandatory. The 1978 text merely permitted states to grant dual protection if they so desired. Through the definition of a "breeder" in Art. 1(c) as including a "person who bred, or discovered and developed, a variety," the 1991 Act makes explicit the requirement that even discovered varieties should be protected.

The 1991 Act recognized the right of breeders to use protected varieties to create new varieties. However, this exception is itself restricted to such new varieties as are not "essentially derived" from protected varieties. The drafters added this restriction to prevent second generation breeders from making merely cosmetic changes to existing varieties in order to claim protection for a new variety. The concept of essential derivation has proved highly controversial in practice, however. Breeders have been unable to agree on a definition of the minimum genetic distance required for second generation varieties to be treated as not essentially derived from an earlier variety and thus outside of the first breeder's control.

From the perspective of farmers, probably the most contentious aspect of the 1991 Act was Art. 15(2) which limited the farmers' privilege to the saving of seed for propagating the product of the harvest they obtained by planting a protected variety "on their own holdings," "within reasonable limits and subject to the safeguarding of the legitimate interests of the breeder." Unlike the 1978 Act, the 1991 version of the farmers' privilege did not authorize farmers to sell or exchange seeds with other farmers. This was criticized as inconsistent with the practices of farmers in many developing nations, where seeds are exchanged for purposes of crop and variety rotation.

It has been suggested that for both social equity and food security reasons there are justifications for providing a farmers privilege for smallholder and resource poor farmers, especially in developing countries, whereby poorer farmers who do not represent an immediate or lucrative market would enjoy the 'farmer privilege' to save seed, while their richer counterparts would be required to pay royalties on saved proprietary seed (Spillane, 1999, 41–42).

A number of developing countries have resisted adopting the 1991 Act as the standard for plant variety protection laws. The foreign ministers of Organization for African Unity issued a statement at a January 1999 meeting calling for a moratorium on IPR protection for plant varieties until an Africa-wide system had been developed that granted greater recognition to the cultivation practices of indigenous communities (Helfer, 2001). This option is not open to those 90 or more countries that have entered into free trade agreements with the United States, since it insists that signatories adopt the 1991 version of UPOV (Drahos, 2002).

The seed saving privilege and the permitted development of non-essentially derived new varieties from protected material were compromises built in to the legislation to take account of public policy concerns. It was appreciated that permitting individuals to privatize food varieties might compromise food security if breeding material was locked up and if farmers were prevented from saving seed for further harvests. However, from the perspective of plant breeders any derivation of new varieties from their protected varieties, whether essential or nonessential, was inconvenient for them and any seed saving by farmers deprived them of new sales. Consequently, they looked to patents law, which does not contain these exceptions, to protect their new varieties.

Plant varieties can be protected in the US under a system of plant patents, or under a system of utility patents or under the Plant Variety Protection Act (PVPA). The Plant Patent Act makes available patent protection to new varieties of asexually reproduced plants. Under this scheme a plant variety must be novel and distinct and the invention, discovery or reproduction of the plant variety must not be obvious. One of the disadvantages of the scheme is that only one claim, covering the plant variety, is permitted in each application. The Federal Circuit Court of Appeals resolved any potential conflict between patent protection and protection under the Plant Variety Protection Act (PVPA) in its decision in *Pioneer Hi-Bred International Inc. v. J.E.M. Ag Supply Inc.* (200 F.3d 1374 (Fed. Cir. 2000), *cert. granted,* 148 L. Ed. 2d 954 (2001). Pioneer's patents covered the manufacture, use, sale, and offer for sale of the company's inbred and hybrid corn seed products as well as certificates of protection under the Plant Variety Protection Act for the same seed-produced varieties of corn. The defendants argued that the enactment of the Plant Variety Protection Act had removed seed-produced plants from the realm of patentable subject matter the Patents Act. The Federal Circuit rejected this argument noting that the Supreme Court held that "when two statutes are capable of co-existence, it is the duty of the courts. .. to regard each as effective".

This was illustrated by *Monsanto Co. v. McFarling* (302F.3d 1291 (Fed. Cir. 2002)) which concerned Monsanto's patent for glyphosate-tolerant plants, the genetically modified seeds for such plants, the specific modified genes, and the method of producing the genetically modified plants. Monsanto required that sellers of the patented seeds obtained from purchasers a "Technology Agreement," in which they agreed that the seeds were to be used "for planting a commercial crop only in a single season" that the purchaser would not "save any crop produced from this seed for replanting, or supply saved seeds to anyone for replanting." Mr. McFarling, a farmer in Mississippi, purchased Roundup Ready soybean seed in 1997 and again in 1998; he signed the Technology Agreement. He saved 1500 bushels of the patented soybeans from his harvest during one season, and instead of selling these soybeans as crop he planted them as seed in the next season. He repeated this activity in the following growing season. This saved seed retained the genetic modifications of the Roundup Ready seed. Mr. McFarling did not dispute that he violated the terms of the Technology Agreement but claimed that the contractual prohibition against using the patented seed to produce new seed for planting, when he produced only enough new seed for his own use the following season, violated the seed saving provision of the PVPA. The Court declined to limit the patent law by reference to the PVPA and Mr. McFarling was found to have infringed Monsanto's patent.

## 5.7 The TRIPS Agreement 1994

Probably the most notorious requirement of the TRIPS Agreement is that in Article 27.3(b), which requires that Members of the World Trade Organization "shall provide for the protection of plant varieties either by patents or by an effective *sui generis* system or by any combination thereof." The principal technical issues concerning the implementation of effective *sui generis* protection of plant varieties under Article 27.3(b) are: (a) what are "plant varieties"? and (b) what *sui generis* options are open to Member states? Article 27.3(b) requires the protection of "plant varieties," but it does not provide a definition of this term. Therefore, national laws have ample room to determine what is to be deemed a plant "variety" for the purposes of protection.

There have been lengthy discussions about the concept of "plant variety," particularly within the framework of UPOV. The scientific notion does not necessarily coincide with the legal concept. The law may require certain characteristics for a *protected* variety that may not be essential for a scientific definition. When breeders seek protection under the traditional plant breeders' rights (PBR) system, plant varieties must meet the criteria that require them to be distinct, uniform and stable (DUS). It has been suggested that "uniformity" and "stability" could be replaced by a criterion of "identifiability," which would allow the inclusion of plant populations that are more heterogenous, and thus take into account the interests of local communities (Seiler, 1998).

The TRIPS Agreement does not prescribe any particular form of protection for plant variety innovations. It could have prescribed the UPOV Convention as the legislative norm, as it did with the Berne Convention for copyrights and the Paris Convention for industrial property. Thus, Members have the option of enacting UPOV-like protection, of including plant varieties within their patent laws, of combining both forms of protection, or of combining UPOV-like protection with biodiversity conservation legislation. The TRIPS Agreement does not prohibit the development of additional protection systems. Nor does it prohibit the protection of additional subject matter to safeguard local knowledge systems or informal innovations, as well as to prevent their illegal appropriation.

A number of developing countries had noted the tension between the development and technology transfer objectives of the TRIPS Agreement and the way in which the Agreement made it possible for rights owners to impose unreasonable terms for technologies. India, noting the difficulties faced by developing countries to obtain access to foreign technology urged that "the TRIPS Agreement may be reviewed to consider ways and means to operationalize the objective and principles in respect of transfer and dissemination of technology to developing countries, particularly the least developed amongst them".[2]

This argument was reflected in part in clause 19 of the Doha Ministerial Declaration of November 2001, which instructed the Council for TRIPS, "in pursuing its work programme including under the review of Article 27.3(b), … [to] be guided by the objectives and principles set out in Articles 7 and 8 of the TRIPS Agreement and shall take fully into account the development dimension." The Doha Ministerial had set the deadline of December 2002 within which the review, referred to in Clause 19 of the Doha Declaration had referred, was to be finalised and reported to the Trade Negotiations Committee (TNC) "for appropriate action". However, after Doha, the discussions in the TRIPS Council were dominated by the consideration of the public health and patenting issue and the question of plant variety protection under Article 27.3(b) was somewhat neglected. However, in anticipation of the Cancun Ministerial, Morocco, on behalf of the African Group of countries made a Joint Communication to the Council for TRIPS, on 20th June 2003, in an endeavour to finalise the longstanding issues relating to the review of Article 27.3(b) (i) indicating the solutions that the African Group considered needed to be found; (ii) setting out possible areas of agreement on issues that have arisen; (iii) providing suggestions on how to resolve issues on which members had not been able to reach a common understanding.[3]

The Joint Communication maintained that the requirement to protect plant varieties should be consistent with and supportive of the public policy goals of Member States relating to food security, nutrition, the elimination of rural poverty, and the

[2] WTO Doc., WT/GC/W/171.

[3] WTO Doc., IP/C/W/404, 20 June 2003.

integrity of local communities. Also asserted was the importance of the preservation of the system of seed saving and exchange as well as selling among farmers in which the legitimate rights of commercial plant breeders should be protected and but balanced against the needs of farmers and local communities, particularly in developing Members.

The Joint Communication urged that in implementing the TRIPS Agreement, the Convention on Biological Diversity and the International Treaty on Plant Genetic Resources in a mutually supportive and consistent manner, Members should retain the right to require, within their domestic laws, the disclosure of sources of any biological material that constitutes some input in the inventions claimed, and proof of benefit sharing.

## 5.8 Patenting of Plant Breeding Methods

A potentially important impact on plant variety protection is a recent development in some patent laws which allow for the patenting of plant breeding methods. For example, in the US a patent has been obtained for the "selective increase of the anticarcinogenic glucosinolates in brassica species" (US Patent 6,340,784, January 22, 2002) and an application published concerning a "method for breeding tomatoes having reduced water content" (US Patent Application 20100095393, April 15, 2010). This raised the possibility that methods of crop breeding to withstanding climate stress can be privatized in the US, which permits so-called methods patents.

The exclusion by the European patent legislation of "essentially biological processes for the production of plants or animals" defined in Article 2.2 of the Biotechnology Directive as consisting "entirely of natural phenomena such as crossing or selection", would have been thought to deny patent protection to plant breeding methods, but this was tested by the European Patent Office (EPO) Enlarged Board of Appeal (EBA) in two determinations, which concerned the two patents which had been granted in the USA for a process involving crossing and selection of broccoli and a process relating to the crossing and selection of tomatoes.

Both of the patent applications were opposed by interested parties. These oppositions were heard by the EPO's Technical Board of Appeal which referred a number of questions to be determined by the EBA.

The EBA ruled that a non-microbiological process for the production of plants which consists of the steps of sexually crossing the whole genomes of plants and of subsequently selecting plants is in principle excluded from patentability as being "essentially biological" within the meaning of Article 53(b) of the European Patent Convention (EPC). The EPC takes account of the 1961 version of UPOV which specifically excludes the patenting of "plant or animal varieties or essentially biological processes for the production of plants or animals'. Rule 23b(5) of the EPC explains that a process for the production of plants and animals is essentially biological if it consists entirely of natural phenomena such as crossing or selection". This language is replicated in the EU Biotechnology Directive which in Article 4.1

excludes from patentability: (a) plant and animal varieties; and (b) essentially biological processes for the production of plants or animals. Article 2.2 states that a process for the production of plants or animals is essentially biological "if it consists entirely of natural phenomena such as crossing or selection."

Such a process does not escape the exclusion of Article 53(b) EPC merely because it contains, as a further step or as part of any of the steps of crossing and selection, a step of a technical nature which serves to enable or assist the performance of the steps of sexually crossing the whole genomes of plants or of subsequently selecting plants.

If, however, such a process contains within the steps of sexually crossing and selecting an additional step of a technical nature, which step by itself introduces a trait into the genome or modifies a trait in the genome of the plant produced, so that the introduction or modification of that trait is not the result of the mixing of the genes of the plants chosen for sexual crossing, then the process is not excluded from patentability under Article 53(b) EPC.

In the context of examining whether such a process is excluded from patentability as being "essentially biological" within the meaning of Article 53(b) EPC, it is not relevant whether a step of a technical nature is a new or known measure, whether it is trivial or a fundamental alteration of a known process, whether it does or could occur in nature or whether the essence of the invention lies in it.

The EBA identified from the jurisprudence the following elements which had been enumerated as relevant to determining whether a process is not essentially biological:

1. The totality of human intervention and its impact on the result achieved is to be determined.
2. This has to be judged on the basis of the essence of the invention.
3. The impact must be decisive.
4. The contribution must go beyond a trivial level.
5. The totality and the sequence of the specified operations must neither occur in nature nor correspond to the classical breeders' processes.
6. The required fundamental alteration of the character of a known process for the production of plants may lie either in the features of the process, i.e. in its constituent parts, or in the special sequence of the process steps, if a multistep process is claimed.[4]

In December 2015, the European Parliament, concerned that the EBA determination might conflict with the EU plant variety legislation adopted a Resolution which asked the European Commission to look into the patentability of products derived from essentially biological processes.[5] Responding to this, on 8 November 2016, the European Commission issued a notice relating to certain articles in the EU Directive

[4] OJ EPO 2016, A27 (G 2/12) and A28 (G 2/13).

[5] P8_TA-PROV(2015)0473: European Parliament Resolution of 17 December 2015 on patents and plant breeders' rights, 2015/2981(RSP).

on biotechnological inventions in which it said that the objective of the legislators in enacting the Biotechnology Directive was to exclude not only biological processes, but also plants or animals obtained from biological processes (EC, 2016). Responding to this Notice, the EPO on 29 June 2017 took a decision to amend the its Regulations in order to exclude from patentability plants and animals exclusively obtained by an essentially biological breeding process (EPO, 2017). Following this amendment in September 2018, the EPO revoked a patent[6] granted in 2013 to Monsanto covering a type of broccoli adapted to make harvesting easier. The revocation follows an opposition filed in 2014 by a coalition of organisations, supported by 75.000 signatures (Kluwer, 2018).

## 5.9 Plant Variety Protection in Developing Countries

From a food security perspective, it should be noted that the UPOV Convention was originally designed to serve the interests of principally European seed breeders and in this respect reflects the industrial interests of European agriculture. Although the TRIPS Agreement does not oblige countries to follow the UPOV model in implementing their plant variety protection obligation in Art.27.3(b) of the TRIPS Agreement, developing countries have tended to adopt legislation on the 1991 UPOV model. As is mentioned above, this model circumscribes the seed-saving possibilities for farmers.

The value of PVRs for encouraging agricultural innovation in developing countries has not been authoritatively established. A UPOV study in 2005 looked at the impact of PVP laws in Argentina, China, Kenya, Poland and the Republic of Korea (UPOV, 2005). It concluded that the impact of PVP varies country-by-country and crop-by-crop. In Argentina, the introduction of new, protected varieties from non-resident breeders was observed in important agricultural crops (e.g. soybean, lucerne) and in horticultural crops (e.g. rose, strawberry). The demand for new, protected varieties was shown by their increased proportion of the certified seed area by 80–90%, particularly, in soybean and wheat. An increase of horizontal cooperation in the seed industry, involving foreign seed companies and agreements for technology transfer between national research institutes and breeding entities with other national companies resulted in more rapid movement of germplasm.

As China's PVP systems have only been in operation for less than 10 years and for a limited number of genera and species and it was not yet possible to evaluate their full impact. Nevertheless, a rapid uptake by farmers of new, protected varieties seen, for example, in maize and wheat in Henan Province was noted, with an increase in the number of breeders in that province, as well as the introduction of new, protected varieties for major staple crops (e.g. rice, maize, wheat),

[6] European patent number 1.597.965.

horticultural crops (e.g. rose, Chinese cabbage, pear), including traditional flowers (e.g. peony, magnolia, camellia) and for forest trees (e.g. poplar).

In Kenya, an increase in the number of varieties developed and released in the 6-year period after the introduction of PVP (1997–2003), compared to the previous 6-year period (1990–1996), across a number of agricultural crops and for maize in particular was noted. Also, the study noted the diversification of the horticultural sector (for example the emergence of the flower industry) and the increased introduction of foreign germplasm in the form of new, protected varieties (especially of horticultural crops).

In the Republic of Korea, a particular impact was the extension of protection to a range of agricultural and horticultural crops, including traditional crops (e.g. ginseng) and varieties of ornamental crops such as rose. The report also noted the stimulation of rice breeding.

A joint project of the Anti-American Institute for Cooperation in Agriculture and the University of Amsterdam carried out in 1994, examined 'the (expected) impact of plant breeders' rights (PBR) on developing countries with respect to: private investment in plant breeding, breeding policies of public institutes, transfer of foreign germplasm, and diffusion of seed among farmers' (Jaffé & van Wijk, 1995). Five countries were used as case studies of which three (Argentina, Chile and Uruguay) had PVP systems already in place, and two (Colombia and Mexico) were about to introduce them. These countries are similar in the sense that there are basically two seed markets. The hybrid seed market is controlled by transnational corporations, whereas the seed market for self-pollinating varieties is dominated by domestic firms. However, Argentina differs from the others in that it is the only country in which PVP right owners have successfully enforced their rights to the extent that their control over seed supply for wheat and soya is comparable to that of their counterparts in the United States. This led the authors of the study report to conclude that in all probability, PVP in that country has 'prevented the local wheat companies from reducing or even terminating their breeding activities and triggered the reactivation of some soya bean breeding programmes'.

In a 2002 study for the UK Commission on Intellectual Property Rights (CIPR), Rajnekar observed that the release of new varieties as an indicator of the impact of PVPs was equivocal evidence as a number of inquiries remain before a conclusive statement on the impact of PBRs on varietal release rates can be accepted as an economic good (CIPR, 2002). First, there is only partial evidence on rates of varietal release in the pre- and post-PVR period. Secondly, the availability of varieties is not necessarily an economic good in itself, as it might be that the increase in varieties may be part of wider appropriation strategies involving planned obsolescence as a means of maintaining market shares, which result in faster rates of varietal turnover and higher varietal release rates. The Final Report of the CIPR noted that the evidence relating to the impact of plant variety protection on research was sparse and mainly from developed countries and indicated that there was little or no evidence that total R&D activity had increased as a result of the introduction of PVP, suggesting that that the main impact of PVP was as a marketing tool (CIPR, 2002).

A UNEP study of 1996 stated that there was "mixed and inconclusive evidence" about the direct benefits of introducing IPRs in plant varieties in developing countries (UNEP, 1996).

Rangnekar concluded that existing evidence of the focus of private sector plant breeding was not entirely promising because "the range of crops focussed on and the type of agro-ecological niches being targeted do not cater to the wider needs of the majority farming populations in developing countries" (Rangnekar, 2002).

No country has yet introduced food security concerns as a factor in implementing plant variety rights protection. However, Kenya, one of the first developing countries to have PVP legislation when it passed the Seeds and Plant Varieties Act, 1975 contains a requirement that 'the agro-ecological value [of the variety] must surpass, in one or more characteristics, that of existing varieties according to results obtained in official tests.' It should be noted however, that there was little demand from domestic breeders for this legislation; it being precipitated more by foreign horticultural firms.

## 5.10 Environmental Impacts of Plant Variety Protection Laws

The environmental and ecological impacts of plant variety laws is also significant from a food security perspective. The high yielding hybrid varieties developed by seed breeding companies has been particularly attractive to farmers. For example, *Zea mays* a maize hybrid had yields 25% greater than standard maize. This was a technological guarantee of future seed sales, as hybrid vigour tended not to be transmitted between generations (see Kingsbury, 2009). However, this had unfortunate side-effects as traditional varieties were replaced by the hybrid causing a loss of genetic diversity. *Zea mays* was susceptible to affliction by the fungus *Helminthosporium maydis* (southern corn leaf blight). By 1970, 80% of the US crop was vulnerable to *H. maydis* and in the wet summer of that year around 20% of the US crop was lost to the blight (Allaby, 2019). A similar loss of genetic diversity has been experienced with the other major cereals (Sofia, 2012; Ahuja & Mohan, 2016).

Another environmental impact of hybridisation is that while it increases the yield in first generation crops, the quality and quantity of subsequent crops deteriorates, and continues to deteriorate, with each replanting. As a consequence, hybridisation operates as a de facto technological protection system obliging farmers to purchase new seed for subsequent plantings (Hubicki & Sherman, 2005). However, hybridisation has only successfully been used in a limited number of crops; it is not used in barley, cotton, millet, rice, soybeans and wheat (Goeschl & Swanson, 2003).

A more effective alternative to hybridization is the genetic manipulation of seed to prevent the germination of any saved seed (see Blakeney, 2004/2005). On 3 March, 1998, the United States Department of Agriculture and the Delta & Pine Land Company were registered as assignees of a US patent "to control plant gene expression".[7] This technology that allowed plant breeders to modify crops so that after germination, they would produce sterile seed. This technology was first known as the "Technology Protection System", after which it has become known as "Genetic Use Restriction Technology" (GURT). There are two types of GURTs: (i) v-GURTs: where the use of a crop variety is controlled through genetically induced seed sterility; and (ii) t-GURTs: where the use of a trait, such as disease resistance or early ripening is controlled. GURTs use "a chemical sensitive genetic switch (responsive, for example, to alcohol or the antibiotic tetracycline) linked to a gene for an enzyme which activates a toxin gene. In the t-GURT system when the toxin gene is switched on, it becomes active in the late stage of seed formation to prevent it germinating (see Daniell, 2002). An advantage of GURTs is that, unlike hybridization, they are applicable to all seeds. As a technological fix it avoids some of the embarrassing publicity from law suits by seed companies against farmers to enforce their intellectual property rights in relation to proprietary seeds, as well as the cost, inconvenience and unpredictability of intellectual property litigation (CFS, 2005; Leahy, 2005).

Pat Mooney, member of the organisation formerly known as Rural Advancement Foundation International (RAFI), now known as the Action Group on Erosion, Technology and Concentration (ETC Group) has coined the term 'Traitor technology' to describe GURTs as a whole and 'Terminator Technology' to describe v-GURTs (Oczek, 2000). This opprobrium is generated, in part, by concerns that if the sterility trait spreads beyond the confines of a field where the GURTs are planted it could produce a 'suicide-plant pandemic' that wipes out an entire species (Mander, 2002). In any event, it is suggested that v-GURT plants could cross-pollinate with non-genetically modified plants, either in the wild or on the fields of farmers who do not adopt the technology. This cross-pollination could reduce yield in the subsequent year due to occurrence of sterile seeds in neighbouring stands This outcrossing is of particular concern where ecological niches and wild relatives exist locally, particularly in the centres of origin of a crop (UNEP/CBD, 2003). An ad hoc technical group meeting of UNEP/CBD also suggested that the application of GURTs might produce low quantities of autotoxic compounds in seeds or other tissues, which may negatively impact non-target organisms (e.g. birds, insects and soil biota).[8] It was also speculated that GURTs might negatively impact the food chain and affect human health due to the additional traits, such as the transfer of allergenicity genes and the transfer of antibiotic resistance (UNEP/CBD, 2003). On the other side of the coin, the National Research Council in the USA has commended v-GURTs as an effective method of confining gene flow (NRC, 2000).

From an intellectual property perspective, it has been suggested that the protection which v-GURT technology assures the seed breeder may go well beyond the time limits of patent and plant variety rights protection (Kariyawasam, 2009), but on the other hand, after the expiry of those rights, the technology is available for reverse

[7] See U.S. Pat. No. 5723765.

[8] *Ibid.*

engineering (Lai, 2014). This is, of course, more likely to be a matter for seed companies than farmers.

## References

Ahuja, M. R., & Mohan, J. S. (Eds.). (2016). *Genetic diversity and erosion in plants: Case histories*. Cham, Switzerland: Springer.
Allaby, M. (2019). *Genetic Erosion in a dictionary of plant sciences*. Oxford, UK: Oxford University Press.
Allard, R. W. (1960). *Principles of plant breeding*. New York: John Wiley & Sons.
Alston, J. M., & Venner, R. J. (2002). The effects of the US Plant variety protection act on wheat genetic improvement. *Research Policy, 31*(4), 527–542.
Blakeney, M. (2004/2005) 'Terminator' technology: A technological fix for an IP problem? Bio-Science Law Review, 7: 229–236.
Butler, L. J., & Marion, B. W. (1985). *The impacts of patent protection on the US seed industry and public plant breeding*. Madison, WI: USA, College of Agricultural and Life Sciences, University of Wisconsin-Madison.
Byrne, N. (1991). *Commentary on the substantive law of the 1991 UPOV convention for the protection of plant varieties*. London: CCLS.
Carew, R., & Devados, S. (2003). Quantifying the contribution of plant breeders rights and transgenic varieties to canola yields: Evidence from Manitoba. *Canadian Journal of Agricultural Economics, 51*(3), 371–395.
CFS (Center for Food Safety). (2005). *Monsanto v US Farmers*. Washington, DC: CFS.
Charnley, B., & Radick, G. (2013). Intellectual property, plant breeding and the making of Mendelian genetics. *Studies in History and Philosophy of Science, 44*, 222–233.
CIPR. (2002). *Integrating intellectual property rights and development policy, report of the commission on intellectual property rights*. London: CIPR.
Cromwell, E., Friis-Hansen, E., & Turner, M. (1992). *The seed sector in developing countries: A framework for performance analysis, ODI* (Working Paper 65), London, ODI.
Daniell, H. (2002). Molecular strategies for gene containment in transgenic crops. *Nature Biotechnology, 20*(6), 581–586.
Drahos, P. (2002). Developing countries and international intellectual property standard-setting. *The Journal of World Intellectual Property, 5*(5), 765–789.
EC. (2016). Commission Notice on certain articles of Directive 98/44/EC of the European Parliament and of the Council on the legal protection of biotechnological inventions C/2016/6997. *OJ C, 411, 3–14.*
EPO. (2017). *EPO clarifies practice in the area of plant and animal patents*. Available at https://www.epo.org/news-issues/news/2017/20170629.html. Accessed 3 Dec 2019.
Fowler, C. (1994). *Unnatural selection: Technology, politics and plant evolution*, Yverdon, Switzerland, Gordon and Breach Science Publishers, 1994.
Goeschl, T., & Swanson, T. (2003). The development impact of genetic use restriction technologies: A forecast based on the hybrid crop experience. *Environment and Development Economics, 8*(1), 149–165.
Gregg, B. R., et al. (1980). Inter-relationships of the essential activities of a stable, efficient seed industry. *Seed Science and Technology, 8*, 202–227.
Harwood, J. (2000). The rediscovery of Mendelism in agricultural context: Erick von Tschermak as plant-breeder. *Comptes rendus de l'Academie des sciences, serie III/Sciences de la vie, 323*, 1061–1067.

Heitz, A. (1991, November 27). *The history of the UPOV convention and the rationale for plant breeders' rights*. Paper delivered at UPOV Seminar on the Nature of and Rationale for the Protection of Plant Varieties under the UPOV Convention, Buenos Aires.
Helfer, L. (2001). *Legal study on intellectual property rights in plant genetic resources*. FAO: Rome.
Hubicki, S., & Sherman, B. (2005). The killing fields: Intellectual property and genetic use restriction technologies. *UNSW Law Journal, 28*(3), 740–757.
Jaffee, S., & Srivastava, J. (1992). *Seed system development: The appropriate roles of the private and public sector*. Washington DC, USA: World Bank.
Jaffee, S., & Srivastava, J. (1994). The roles of the private and public sector in enhancing the performance of seed systems. *The World Bank Research Observer, 9*(1), 97–117.
Jaffé, W., & van Wijk, J. (1995). *The impact of plant breeders' rights in developing countries: Debate and experience in Argentina, Chile, Colombia, Mexico and Uruguay*. The Hague, The Netherlands: Directorate General International Cooperation, Ministry of Foreign Affairs.
Kariyawasam, K. (2009). Terminator technology as a technological means of forcing intellectual property rights in plant Germplasm: It's implications for world agriculture'. *European Intellectual Property Review, 31*(1), 37–44.
Kevles, D. (2011). New blood, new fruits: Protection for breeders and originators, 1789–1930. In M. Biagoli, P. Jazi, & M. Woodmansee (Eds.), *The making and unmaking of intellectual property: Creative productions in legal and cultural perspective* (pp. 253–267). Chicago/London: University of Chicago Press.
Kingsbury, N. (2009). *Hybrid: The history and science of plant breeding*. Chicago: The University of Chicago Press.
Kloppenburg, J. R. (2004). *First the seed: The political economy of plant biotechnology, 1492–2000* (2nd ed.). Madison, WI: University of Wisconsin Press.
Kluwer. (2018). *EPO revokes Bayer broccoli patent*, Kluwer Patent Blog. Available at http://patentblog.kluweriplaw.com/2018/11/16/epo-revokes-bayer-broccoli-patent/. Accessed 3 Dec 2019.
Knudson, M. (1990). The role of the public sector in applied breeding R&D. *Food Policy, 15*, 209–217.
Kolady, D. E., & Lesser, W. (2009). Does plant variety protection contribute to crop productivity? Lessons for developing Countries from US wheat breeding. *Journal of World Intellectual Property, 12*(2), 137–151.
Lai, J. C. (2014). The exhaustion doctrine and genetic use restriction technologies: Look at Bowman v Monsanto. *Journal of World Intellectual Property, 17*(5–6), 129–141.
Leahy, S. (2005). *Monsanto "Seed Police" scrutinize farmers*. Available at https://corpwatch.org/article/us-monsanto-seed-police-scrutinise-farmers. Accessed 1 Dec 2019.
Leskien, D., & Flitner, M. *Intellectual property rights and plant genetic resources: Options for a sui generis system* (Issues in Genetic Resources No.6). Washington, DC: IPGRI.
Lesser, W. H. (1990). Sector issues II: Seeds and plants. In W. E. Siebeck, R. E. Evenson, W. H. Lesser, & C. A. Primo Braga (Eds.), *Strengthening protection of intellectual property in developing countries* (pp. 59–68). Washington, DC: The World Bank.
Lesser, W. H. (1991). *Equitable patent protection in the developing world: Issues and approaches*. Tskuba, Japan: Eubios Ethics Institute.
MacLeod, C. (1996). Concepts of invention and the patent controversy in Victorian Britain. In R. Fox (Ed.), *Technological change, methods and themes in the history of technology, Abingdon* (pp. 137–154). Oxford: Routledge.
Mander, J. (2002). Machine logic: Industrialising nature and agriculture. In A. Kimbrell (Ed.), *The fatal harvest reader: The tragedy of industrial agriculture* (pp. 17–23). San Francisco: Foundation for Deep Ecology.
Morris, M. L. (Ed.). (1998). *Maize seed industry in developing countries*. Boulder, CO/London: Lynne Rienner.
Murphy, D. J. (2007). Origins of plant breeding. In D. J. Murphy (Ed.), *Plant breeding and biotechnology: Societal context and the future of agriculture* (pp. 9–22). Cambridge: CUP.

NRC. (2000). *Genetically modified Pest-protected plants: Science and regulation*. Washington, DC: National Academy Press.

Oczek, J. P. (2000). In the aftermath of the "terminator" technology controversy: Intellectual property protections for genetically engineered seeds and the rights to save and replant seed. *Boston College Law Review, 41*, 627–657s.

OECD. (2018). *Concentration in seed markets. Potential effects and policy responses*. Paris: OECD.

Palladino, P. (1994). Wizards and devotees: On the Mendelian theory of inheritance and the professionalization of agricultural science in Great Britain and the United States, 1880–1930. *History of Science, 4*, 409–444.

Palladino, P. (2002). *Plants, patents and the historian: (Re)membering in the age of genetic engineering*. Manchester, UK: Manchester University Press.

Perrin, R. K., Kunnings, K. A., & Ihnen, L. A. (1983). *Some economic effects of the US Plant Variety Protection Act of 1970*. Raleigh, NC: North Carolina State University.

Pottage, A., & Sherman, B. (2007). Organisms and manufactures. On the history of plant inventions. *Melbourne University Law Review, 31*, 539–568.

Raeburn, P. (1995). *The last harvest. The genetic gamble that threatens to destroy American agriculture*. New York: Simon and Schuster.

Ramaswami, B. (2000). *Intellectual property rights in agriculture: implications and challenges for public research and policy*. New Delhi, India: Indian Statistical Institute.

Rangnekar, D. (2000). *Intellectual property rights and agriculture: An analysis of the economic impact of plant breeders' rights*. London: Actionaid.

Rangnekar, D. (2001). *Access to genetic resources, gene-based inventions and agriculture* (Commission on Intellectual Property Rights (CIPR) Study Paper 3a). London: CIPR.

Rangnekar, D. (2002). R&D appropriability and planned obsolescence: Empirical evidence from wheat breeding in the UK (1960–1995). *Industrial and Corporate Change, 11*(5), 1011–1029.

Rossman, J. (1935). The preparation and prosecution of plant patent applications. *Journal of the Patent Office Society, 17*, 632–645.

Sanderson, J. (2017). *Plants, people and practices. The nature and history of the UPOV convention*. Cambridge: Cambridge University Press.

Seiler, A. (1998). Sui generis systems: Obligations and options for developing countries. *Biotechnology and Development Monitor, 34*, 2–5.

Silva Dias, J., & Ryder, E. J. (2012). The impact of plant breeding on the world vegetable industry. *Acta Horticulturae, 935*, 13–22.

Spillane, C. (1999). *Recent developments in biotechnology as they relate to plant genetic resources for food and agriculture* (Background Study Paper No. 9). Rome: Commission on Genetic Resources for Food and Agriculture.

ten Kate, K., & Laird, S. A. (1999). *The commercial use of biodiversity: Access to genetic resources and benefit sharing*. London: Earthscan.

UNEP. (1996). *The impact of intellectual property rights system on the conservation and sustainable use of biological diversity and on the equitable sharing of benefits from its use*. UNEP: Nairobi, Kenya.

UNEP/CBD. (2003). *Report of the Ad Hoc technical expert group meeting on the potential impacts of genetic use restriction technologies on smallholder farmers*, Indigenous and Local Communities and Farmers' Rights. UNEP/CBD/SBSTTA/9/INF/6.

UPOV. (1978). Amendments to facilitate the joining of the Union by Further States, UPOV doc. DC/PCD/1.

UPOV. (1987). *The history of plant variety protection, in the first twenty-five years of the international convention of the protection of new varieties of plants*. Geneva, Switzerland: UPOV.

UPOV. (2005). *Report on the impact of plant variety protection*. Geneva, Switzerland: UPOV.

Williams, S. B. (1983). Intellectual property aspects of plant variety genetic engineering: View of an American lawyer. *Plant Variety Protection*, 13–36.

# Chapter 6
# Trait Breeding and Plant Varietal Rights

**Seyed Hossein Jamali**

**Abstract** This chapter looks at the relationship between plant breeding and trait breeding. It details the problem of variety ownership boundaries and the concept of essentially derived varieties (EDV). The chapter looks at the impact of patented accelerated breeding methods. It looks at the technical distinctness, uniformity, and stability (DUS) requirements of the UPOV system which largely rely on morphological (phenotypic) traits irrespective of their value for cultivation and use of a newly bred variety. The chapter examines the Australian approach for resolving varietal disputes and concludes with an examination of the challenges of genome editing technologies.

**Keywords** Trait breeding · UPOV requirements · Essentially derived variety · Accelerated breeding · Genome editing

## 6.1 Introduction

The current forms of intellectual property rights for protecting plant varieties substantially differ in second breeders' access to protected germplasm as breeding material. While the patent regime vests in the licensees an exclusive right of restraining any derivation from patented variety, the International Union for the Protection of New Varieties of Plants (UPOV) convention recognises second breeder's exemption on exploiting a protected variety for further breeding. However, the latest Act (1991) of the UPOV Convention restricted freedom-to-operate of second breeders through introduction of 'essential derivation' concept. The key driver for imposing such circumscription was low distinctness standard and limited infringement provisions in earlier versions of UPOV convention. The coincidence of these drawbacks with advent of biotechnological tools (considerably genetic engineering) raised concerns of plant breeder's rights (PBRs) holders over plagiaristic practice. It was

S. H. Jamali (✉)
Independent Researcher, Melbourne, Australia



M. Blakeney, K. H. M. Siddique (eds.), *Local Knowledge, Intellectual Property and Agricultural Innovation*, https://doi.org/10.1007/978-981-15-4611-2_6

argued that genome of derived varieties via such trait breeding tools (discussed below) are a copy of genetic (and hence phenotypic) makeup of their founder protected varieties whilst harbouring minor modifications or a cosmetic change. Therefore, second breeders have to develop independently derived varieties, otherwise, commercialization of essentially derived varieties (EDVs) are under scope of initial PBRs.

Article 14(5)(b) of UPOV convention consists an intricate mixture of technicalities relevant to 'distinctness' and 'conformity' in 'essential characteristics', as well as 'predominant derivation' practice for a fulfilled definition of EDVs.[1] Further, UPOV's explanatory notes narrows down clear distinctness level of an EDV to just one or a few morphologically traits from its founder initial variety.[2] Accordingly, another standard (conformity) has been added to the existing one (distinctness), where plant variety protection (PVP) system had been relying on from the outset for technical examination of candidate varieties. In other words, conformity (similarity) and distinctness (dissimilarity) assessments are two side of the same coin. Whilst the distinctness criterion could be satisfied by, at least, one clear and stable morphological (phenotypic) trait between variety pairs, assessment of varietal relatedness or conformity demands a holistic approach. Therefore, making decision on conformity of an EDV to its founder initial variety is not as facile as distinctness testing. From scientific perspective, a more and evenly-distributed sampling of genome of two disputed varieties provides an unbiased estimation of varietal relatedness. This could be satisfied by DNA markers which in reference to morphological traits, are much more abundant and could be chosen in such a way that cover the whole genome. In this way, Heckenberger suggests a redefinition of the term 'essential characteristics' (Heckenberger, 2004, 96) by which molecular data can also play the same role as conventional, phenotypic traits exemplified in UPOV's guidance on EDVs (UPOV, 2017, 5).

So far, salient interpretations of essential derivation concept have been diverse: Lesser and Mutschler regard it as a totally unworkable system, if measured and identified by species-specific criteria rather than a trait-based approach (Lesser & Mutschler, 2004, 1119); Janis and Smith believe that measurement of conformity within UPOV's PVP system is irrelevant, as a genotype-centred approach for measurement and identification of EDVs is, at least in long-term perspective, incompatible with phenotype-centred technicalities of PVP system (Janis & Smith, 2007, 1561); and finally, Sanderson concludes that looking at 'cultural and practical' value of the distinctive trait(s) in a putative EDV variety is of much utmost importance (Sanderson, 2017, 224–27), as reflected in Section 4 of Australia's PBR Act. This adds another dimension to the hybrid nature (technical and legal) of essential derivation concept for identification and examination of EDVs.

---

[1] UPOV convention, Act of 1991, Article 14(5)(b).

[2] Explanatory notes on essentially derived varieties under the 1991 Act of the UPOV convention (2017), https://www.upov.int/edocs/expndocs/en/upov_exn_edv.pdf.

In this chapter, I analyse freedom-to-operate of second breeders within the purview of trait breeding. For background, the general objective of trait breeding (or trait improvement) is to alleviate flaw(s) of an already improved variety, take for example transferring a foreign resistance gene (via genetic engineering) to an improved, but, pest susceptible variety. In this way, we can also speak of random mutagenesis and backcrossing as conventional trait breeding methods which have been used for this purpose, as well. In contrast, plant breeding aims at developing a new variety, where for example through classical bi-parental cross breeding, two diverse parents are hybridised then desired progenies are selected through successive segregating generations. As will be discussed later, plant and trait breeding would also play a role in strengthening or attenuating plant varietal rights.

## 6.2 A Separation Line Between Plant and Trait Breeding

The convention drafters unconsciously drew a separation line between plant and trait breeding by highlighting methods, namely selection of a (naturally-occurred, induced, or somaclonal) mutant plant (collectively mutation breeding), backcrossing or genetic engineering,[3] that probably give rise to development of EDVs.

The relation between the two latter methods, which are widely used by breeders, and development of EDVs could be evaluated. While backcrossing is an independent trait breeding method for transferring a gene of interest, for example resistance to a pest, from a donor variety to a susceptible one, genetic engineering could not be independently deployed for variety improvement. At least in agricultural crops, a variety is firstly utilised for genetic transformation, as in comparison to other varieties, it demonstrates adequate competency to regeneration from transformed cells and overall genetic engineering process. Due to random nature of transgene integration into genome during each transformation process, the resultant transgenic plant is a unique event that is selected, from a large amount of putatively transformed explants on the basis of harbouring merely one copy of the transgene and not inactivating any endogenous gene at insertion site. Thereafter, the selected event is used as a transgene donor for improvement of numerous varieties, where each is agronomically adapted to a specific growing environment. There is no indication so far that transgenic varieties are protected by PBRs. Instead, breeding companies are of most interest in patenting of transgenic plants to protect their innovative work. This could be shown by patenting of more than 1,400 soybean cultivars and varieties (as of January 2020) in the USA that are made resistant to glyphosate weedkiller through backcrossing to A3244, the recipient parental line of transgenic event MON89788, or subsequently derived transgenic varieties therefrom.[4]

[3] UPOV convention, Act of 1991, Article 14(5)(c).

[4] The figure was obtained by counting citations to US Patent 7,632,985.

In the case of backcrossing, the ultimate objective of breeder is to revive genome of recipient variety together with the trait of interest (transferred from donor variety) after multiple generations. Theoretically, in each backcrossing generation half of the genome of recipient (known as recurrent) variety is restored. Thus, it is expected that after six backcross generations the recipient variety has gained almost all (99%) of its genome. In other words, in case of genetic engineering process, almost all DNA (except the integrated transgene in a non-coding part of genome) from the plant harbouring transgenic event is removed.

## 6.3 Variety Ownership Boundaries

The scientific community has tried to define quantitative boundaries around protected varieties. In general, a broad set border erodes second breeders' rights (as it would give rise to more claims of EDVs by initial breeders), whilst a narrow one jeopardise initial breeders (that leads to more claim of independently derived varieties by second breeders). Noli, Teriaca and Conti reviewed and elaborated on three strategies, suggested by Van Eeuwijk and Law (2004), for establishing similarity thresholds between initial and essentially derived variety pairs (Noli et al., 2013). However, until now crop-specific guidelines for resolving varietal disputes limit to a handful of plants, namely perennial ryegrass, maize, oilseed rape, cotton and lettuce, as adopted by International Seed Federation.[5] Yet, UPOV's technical standards for examining distinctness of candidate varieties from common knowledge varieties cover a total of 330 species(as of January 2020).

While distinctness standard is based on morphological traits (those that are less or unaffected by environmental factors), assessment of conformity between the initial variety and putative EDV mostly rely on similarity coefficients obtained from molecular data. However, technological changes towards genotyping platforms with more abundance and genome coverage opens up more space for freedom-to-operate of second breeders. For instance, the threshold for indisputable essential derivation (red) zone increased from 90% (based on multi-allelic simple sequence repeats markers) to 95% (based on bi-allelic single nucleotide polymorphism markers) aimed at determining conformity of maize inbred lines (ISF, 2014). In this way, updated thresholds are promising for relieving the onus from second breeders, as a maize inbred line that resembles 90% to its founder variety (based on simple sequence markers) were hitherto on the border of uncertainty (orange) and non-distinctness (red) zones.

Lesser and Mutschler's (2004) analysis of 'essential derivation' recognizes it as an unworkable system and finds protection of discrete traits (those inherited by single genes) a more concerning issue than the matter of derivation. They state "[T]he existence of the trait in varieties for which the initial variety was included in

[5] https://www.worldseed.org/our-work/trade-rules/#essential-derivation.

the breeding ancestry would be prime face evidence of essential derivation, with no need to establish a proportion of genetic material" (Lesser & Mutschler, 2004, 1119). Otherwise, only a low set EDV threshold like 75% similarity (*i.e.* a broad set border around protected variety) would prevent second breeders from unauthorised utilisation of protected varieties, whilst maintaining incentives for initial breeders to continuously invest on lengthy pre-breeding or germplasm enhancement programs, whereby desirable traits are introgressed from crop wild relatives to elite germplasm. Likewise, some breeders of vegetatively-propagated plants advocate for adopting a broader scope of essential derivation to identify all mutant and genetically-modified varieties as EDVs. In their view, any variety derived from a single parent through trait breeding approaches is an EDV regardless of differences it has from its founder variety. Thus, narrowing of EDV definition to one or a few distinctive traits (UPOV, 2017, 6) (e.g. a new petal colour or resistance to a pathogen) leaves no incentive for investment in future breeding programs.

## 6.4 Faded Separation Line

A recent large-scale study (Fradgley et al., 2019) on pedigree of wheat genotypes taken from 38 countries debunks that essential derivation is a method-driven concept. Generally, it is assumed that classical cross breeding leads to new allelic and phenotypic combinations. However, comparison of computer simulations (in the absence of selection) with observed genotypic (marker) data surprisingly showed that "wheat varieties derived from biparental crosses commonly share over 80% of their genetic material with one parent, which is greater than would be expected by backcrossing to a recurrent parent" (Fradgley et al., 2019, 11). It is inferred that the underlying reason is breeder's strong selection for desirable agronomic, quality, and resistance traits in segregating genotypes favouring the superior parent (Fradgley et al., 2019, 11–12). These findings provide a fresh perspective on essential derivation concept and paves way to research on other plants.

## 6.5 Accelerated Pre-breeding and Patents

It is known that crop wild relatives are valuable source of agronomically-important genes. However, the genetic diversity was eroded roughly 10,000 years ago through domestication process, and subsequently modern plant breeding during the last century (Tanksley & McCouch, 1997, 1063). The percentage of genetic diversity not transferred from wild progenitors differ between crops (Smith et al., 2015, 66–67), with the hefty deprivation (69%) occurring in cultivated wheat (Haudry, et al., 2007, 1512). The resultant narrow genetic base function as a bottleneck against improving varieties in the face of climate change and rapid evolving pathogens. Hajjar and Hodgkin (2007) surveyed a 20-year period pertaining contribution of crop wild relatives in providing useful traits for 16 mandate crops of the Consultative Group on International Agricultural Research (CGIAR). The study demonstrates significant contribution of biotic resistance in reference to other traits like abiotic stress, yield, quality and male sterility or fertility restoration. In this way, single-gene inheritance of most biotic resistance genes makes it feasible to enhance elite germplasm in a much straightforward way than yield and tolerance to biotic stresses which are multigene-based and hence less heritable. The process entails considerable risk, time and cost, as there would be barriers to inter-specific hybridisation and, if successful, the resultant progeny should undergo repeated backcrossing to recipient variety for removing associated undesirable loci (linkage drag) associated with the introgressed gene.

A recent innovative gene cloning method allows for accelerated germplasm enhancement in much faster pace than conventional procedure (Arora et al., 2019, 139). The inventors have filed patent application for SrTA1662, one of the four stem rust resistance genes that were harvested from goat grass (the precursor species and donor of D genome to bread wheat), as well as the method[6]. It is suggested that the cloned genes could be stacked into susceptible varieties by genetic engineering. Whether it is a new indication of patenting of genetic resources that would rekindle debates on return of derived products from freely-accessed materials to their center of origin under any form of proprietary rights, the novel pre-breeding method would pave way for further circumscription of second breeders. In this way, patent regime offers a stronger protection than PVP system to restrain second breeders from transfer of patented trait to their susceptible breeding materials, particularly where patenting of plant varieties is not allowed. In this way, plant varieties are no longer free entities for further breeding during the term of PVP, as inclusion of 'limited breeder's exemption' in national patent laws of some European countries, mandates elimination of patented trait(s) in the derived varieties unless second breeders obtain license from the patent holder (Prifti, 2017, 112).

## 6.6 Intertwined Technicalities of Interrelated Legislations

The technical distinctness, uniformity, and stability (DUS) requirements of the UPOV system largely rely on morphological (phenotypic) traits irrespective of what the value for cultivation and use (VCU) of the newly bred variety would be. Accordingly a new variety, irrespective of being independently or essentially derived from an initial protected variety, should satisfy clear and stable morphological distinctness (D within DUS) to "[A]ny other variety whose existence is a matter of common knowledge" and have sufficient uniformity (U) and stability (S) in

[6] Patent applications WO2019140351 and WO2019138244 for the cloned gene and the method, respectively.

expression of traits through successive generations.[7] Yet, in practice, comparison of a candidate variety with all common knowledge varieties is infeasible. Instead, a convenient subset (colloquially known as reference collection) is used by DUS examiners.

Also, in Europe and many other countries, statutory variety registration is compulsory before variety release and marketing. Same DUS standards are applied as a solid base for listing new varieties of agricultural crops, vegetables, and fruits.[8] Variety descriptions, as the outcome of DUS testing, are interchangeable between PVP and registration procedures. Therefore, in these countries, a protected variety (except ornamental plants) is a listed one, as well. However, distinction (D within DUS) of candidate varieties for protection may be fulfilled through comparison with common knowledge varieties, while for listing purposes (variety registration) a panel of already registered varieties in national lists would be sufficient for examination (Van Wijk & Louwaars, 2014, 49). Variety registration also mandates crop varieties to demonstrate superior VCU over check (control) varieties at multi-environmental trials.[9] In this way, distinctness testing may be a barrier to commercialization of crop varieties with sufficient VCU over existing varieties. For example, some records have stipulated that a quarter of alfalfa (*Medicago sativa* L. ssp. *sativa*) Annicchiarico et al., 2016, 2) with sufficient VCU were not registered owing to lack of distinctness from existing varieties. This situation explains the fact that release of improved varieties, particularly of agricultural crops, are totally based on statistical outperformance over check varieties in term of yield and/or other VCU traits. But, DUS testing has come to a standstill in distinguishing newly bred from existing varieties, as required by both PVP and variety registration systems. In simple words, improved varieties that express significant performance over existing varieties may not necessarily be distinct in characteristics that are used for variety registration. The bottleneck is apparent in species with narrow genetic base (e.g. soybean) or where quantitative DUS traits (that does not provide a consistent distinctness across different environments) outnumber qualitative ones that are more suitable for fulfilment of a precise distinction (of e.g. perennial forages).

In recent years, the UPOV has also recognized a complementary role for integration of DNA markers into DUS testing via two models (UPOV, 2013). In contrast to morphologically-based DUS traits, DNA markers are much more abundant, free from environmental effects, and could be applied to seed or early stages of plant growth, hence providing a fast and unbiased outcome for registration. According to the first model, a combination of morphological and DNA-based distances is used to select only the most similar comparable varieties (from the reference collection) to candidates for entering into side-by-side field (or glasshouse) comparisons. In few crops like maize (UPOV, 2014) and soybean (UPOV, 2018), this approach

---

[7] UPOV convention, Act of 1991, Articles 7–9.

[8] https://ec.europa.eu/food/plant/plant_propagation_material/plant_variety_catalogues_databases_en.

[9] Commission Directive 2003/90/EC: Rules on minimum characteristics and minimum conditions for examining certain varieties of agricultural plant species.

has led to considerable reduction in workload and hence costs of DUS trials. In model 2, those diagnostic DNA markers are envisaged that are tightly linked to or derived from the underlying causative polymorphism(s) controlling the effect of genetic locus of a given DUS trait. So far, practical examples limit to molecular assays that predict seasonal growth habit (winter/spring type) (Cockram, et al., 2009) and 15 other DUS traits (Cockram et al., 2012) in barley, as well as diagnostic ones for disease resistance traits in tomato (Arens et al., 2010). The key advantages of these 'characteristic-specific molecular markers' lie in alleviating necessities for implementing special trials (e.g. seasonal growth habit) or conducting difficult, time-consuming and costly tests (e.g. disease resistance). Despite this, a search in literature finds expeditiousness of integrating DNA profiles into DUS examinations, though, majority of markers used so far does not necessarily associate with genomic regions (Jamali, et al., 2019). Using markers with unknown function contradicts with International Seed Federation's view, as they would jeopardise PBRs by decreasing minimum distance between varieties to just one base pair of neutral difference (ISF, 2012, 12).

## 6.7 Australian Approach for Resolving Varietal Disputes

As a unique approach, plant breeder's rights Act (1994) of Australia defines a role for national authority as responsible for defining essential characteristics and administering EDVs.[10] Thus, before any litigation, the national authority determines whether a variety is independently or essentially derived from an initial variety. Recognizing that situation, an EDV claim by initial breeder could be rebut if the second breeder exhibits important (as distinct from cosmetic) features of distinguishing trait(s) in the derived variety.[11] This could be translated to a new merit or VCU in terms of yield, agronomic traits (e.g. earliness in maturity), quality (e.g. increased levels of unsaturated fatty acids in grain), and resistance to biotic (pests or pathogenic fungi) and abiotic (e.g. drought) stresses. This approach recognises VCU traits for identifying EDVs[12], however, defining such attributes to a distinctive trait would be species-specific. While a different colour in awns of a derived variety gives no justifiable VCU to cereal crops (e.g. barley), it confers an aesthetic value to an ornamental plant species. In some cases, second breeders may take advantage of auxiliary function of traits to relieve from any EDV claim. For example, smaller lodicule size that confers cleistogamous (non-opening) flowers and leads to complete (100%) self-pollination in barley may be considered as an insignificant characteristic, but it can be accepted as a VCU trait because it also delivers a means of escape from flower-borne head blight disease (Nair et al., 2010, 490).

[10] Plant Breeder's Rights Act 1994 (Cth), Articles 40-41.

[11] Id. Article 4(c).

[12] Id. Article 3, Definitions: essential characteristics.

## 6.8 Genome Editing and Challenges

Genome editing is a site-directed mutagenesis tool that enables breeders precisely and predictably manipulate base pairs of endogenous genes (or their regulatory elements) as well as introduction, deletion or replacement of large DNA fragments (Chen et al., 2019). The technology is synonymous with gene editing, genome engineering, and precision breeding in scientific literature. A few various tools have been devised for genome editing, however, the most prospective one is the 'clustered regularly interspaced short palindromic repeat' (CRISPR) system owing to its operational simplicity and flexibility. Regardless of growing patent landscape for genome editing methods (Schinkel & Schillberg, 2016) and resultant edited traits, one can assume substantial challenges presented by two main attributes of this technology *i.e.* precision and speed of breeding to PVP system.

## 6.9 Precision Drawbacks

Gene-edited varieties are largely tailored for a VCU trait, where the PVP system ignores them for technical examination of candidate varieties. Generally, VCU traits (e.g. yield) are not included in DUS test guidelines, as they significantly interact with environment and hence hinder precise distinction and identification of varieties. Even if it is the case, DUS and VCU trials are mutually exclusive. The layout of DUS trials are designed in such a way that they provide optimal conditions for growth of plant varieties, as required for expression of phenotypic-based DUS traits. By contrast, VCU trials are conducted according to actual farming practices to be a good representative of agronomic conditions in the field. Therefore, it is necessary to design special trials for assessment of VCU traits. This could be illustrated by developing a wheat plant with tolerance to pre-harvest sprouting, a crucial trait at harvest time in wet area, through knocking out endogenous *TaQsd1* gene (Abe et al., 2019). In this way, registering the gene-edited wheat that differs from its founder variety in merely three nucleotides (one point mutation per A, B, and D sub-genomes), while indistinguishable for other characteristics, requires special conditions to be met for expression of pre-harvest sprouting. Likewise, preparing analogous situation would be costly for expression of stacked herbicide tolerance in a co-edited wheat variety (Zhang et al., 2019).

Genome editing would also increase varietal disputes between breeders owing to reduction of distinctness level to just a few nucleotides, provided that the founder variety be protected by PBRs. Yet, the Australian approach, that links DUS to VCU, would be a straightforward approach for settling such ownership claims.

## 6.10 Diminution of Lead Time

Another feature of genome editing lies in the speed of delivering improved plant varieties to farmers and consumers. Chen and colleagues demonstrate that genome editing (by taking 4–6 years) outpaces other trait breeding methods *i.e.* mutation breeding (8–10 years), transgene breeding (8–12 years), and backcrossing (8–10 years) (Chen et al., 2019). The technology could also be combined with speed breeding protocols that permit growing up to six generations per year for cereals and pulses, and four generations for canola through growing plants under extended photoperiod (22 h) and controlled temperature conditions (Watson et al., 2018). Combination of these tools deliberately erodes the lead time (period between the beginning of breeding program and release of resultant varieties) that initial breeders hitherto enjoyed as an opportunity to recoup research and development costs, via collection of royalties, before a second breeders develop an EDV (Janis & Smith, 2007). In order to remedy the imbalance, some reforms to breeders' exception has been suggested. For example, a phased-in breeders' exception restrains second breeders derive any new variety for X years (depending on crop), whereas allows for utilising protected variety for the remaining years of PVP term (ASTA, 2004).

## 6.11 Self-Regulatory Protection Systems

Discovery of heterosis phenomenon in maize, with the outcome of outperforming F1 progeny to homozygous parents, was a landmark achievement in plant breeding history. In the meantime, breeders have also exploited hybrid technology as a 'self-regulatory protection' spin-off, so as some companies do not seek protection of their hybrid varieties by PBRs. This system which is also known as 'biological protection', obliges farmers to purchase seeds for every growing season that otherwise seed saving therefrom brings a remarkable reduction in yield and uniformity due to segregation of heterozygous loci in successive generations. The hybrid technology has been expanded to a variety of crops and vegetables using artificial hand pollination or deployment of pollination control technologies (Chen & Liu, 2014).

In 1996, a patent entitlement jointly to Delta & Pine Land and US Department of Agriculture opened a door for engineered biological protection by controlling gene expression in plants. Since then, over 40 patent families have been granted to what is literally termed 'genetic use restriction technologies' (GURTs) (Lombardo, 2014). There are various mechanisms of GURTs, though, they are categorised under two levels. At variety level (V-GURT), a cytotoxic protein will be expressed by a lethal transgene during late embryogenesis to result in downgrading to sterile seeds. As a consequence, the harvested grain could be just used as food but not seeds. This certainly stymies replication of patented entity (trait/variety) through successive generations, as in cases of self-pollinating crops (e.g. soybean) which breed true, it is hard for patent owners to enforce their rights over farm-saved seeds. Transgenic sterile seeds were never marketed as they faced with strong oppositions at international level, particularly by Canadian-based nongovernment organization Action Group on Erosion, Technology and Concentration which coined them collectively as 'terminator', 'traitor' or 'zombie' technologies (Lombardo, 2014). However, concern over release of sterile seeds is at least maintained in India, where registration applicants have to pledge that the candidate variety does not contain any gene(s) of terminator technology.[13] In another regulatory mechanism (T-GURT), an improved trait could be switched off (protected), unless the seeds are treated by a chemical inducer that behaves as a stimulus for trait expression during plant growth phase. The T-GURT technology can also be an efficient measure in impeding movement of transgenic traits to other varieties through cross-pollination (Hills et al., 2007), hence safeguarding the rights of trait owners.

The future role of trait breeding is also considerable in attenuating self-regulatory protection system. Recently, it was shown that the heterozygosity and the consequent heterosis vigour could be fixed in hybrid rice. The inventors, co-edited three endogenous genes to induce *Mitosis* instead of *Meiosis* in reproductive cells and then eliminated doubled chromosomes of male gamete by knock down of another gene to produce progenies that are genetically identical to the hybrid parent (Wang et al., 2019). Therefore, it is conceivable for seed industry to constantly clone heterosis vigour of a hybrid variety without prerequisite for crossing parental inbred lines every year, however, the devised method has been protected by patent[14] rights in China. In another approach which is based on silencing involved genes in recombination during meiosis, homozygous parental lines of a hybrid seed are reproduced to reconstitute the original heterozygote genotype (Dirks et al., 2009). Whether the 'reverse breeding' techniques is a portent for breeding companies, it allows for unauthorised appropriation of parental lines protected by intellectual property rights (patent or PVP) or trade secrets.

## 6.12 Conclusion

Since introduction in 1991, the essential derivation has been the subject of contentious debate among breeders. Some do not tolerate UPOV's narrowing definition *i.e.* confining EDVs to one or a few distinctive traits from initial variety, rather are in favour of adopting a broader scope that covers trait breeding methods as *prima facie* cause of essential derivation. This means that, at least in vegetatively-propagated plants, all (natural, induced, or gene-edited) mutant and transgenic varieties derived from an initial protected variety should be considered as EDVs. From another perspective, providing the VCU of distinctive trait(s) can rebut an EDV claim by initial

[13] The Protection of Plant Varieties and Farmers' Rights Act (2001) of India, Article 18(c).

[14] CN 110,257,418.

breeder. However, implementation of this approach, as reflected in Australia's PBR Act, would be challenging in vegetables, fruits, and ornamental crops, where phenotypic appearance of varieties has generally a marketing value. For example, initial breeder of an eggplant variety would consider any variation in fruit skin colour of an EDV from his leading variety in the market as a 'cosmetic' change. Nevertheless, the testing authority may accept it as an added aesthetic value to initial variety that is culturally accepted within consumer society. In addition, considering sustainability attributes of varieties to VCU criteria, as adopted by France in VCUS testing, would be helpful. In this way, down-regulation of polyphenol oxidase in apple will no longer be cosmetic, since the resultant non-browning fruits have impact on food security through minimising food waste. Therefore, interpretation of Article 4(c) of Australia's PBR Act: "it does not exhibit any important (as distinct from cosmetic) features that differentiate it from that other variety" depends on where to place the border of important feature and cosmetic attribute. On the other hand, implementation of essential derivation concept is unfeasible in certain plant species. For example, plant varieties of species with narrow genetic base are morphologically similar, so as varietal differences have come down to merely VCU traits. Conversely, fulfilment of phenotypic-based DUS criteria in population varieties of genetically heterogeneous species is even a challenge that makes conformity assessment irrelevant.

Latest UPOV's seminar on "the impact of policy on essentially derived varieties (EDVs) on breeding strategy" reflects a disparity between the essential derivation concept and breeder's practice[15]. This notion was examined in present paper with the inference that backcrossing and genetic engineering, as the most trait breeding methods of choice by breeders, unlikely give rise to development of EDVs. The seminar also highlights concerns of breeders over genome editing that create varieties with diminutive difference at one or a few nucleotides. It should be noted that genome editing follows a similar procedure to genetic engineering: an edited line is used as a donor for improvement of other varieties, where each is agronomically adapted to a specific growing environment. For example, Collectis SA describes patented soybean plants embodying FAD2-1A and FAD-1B knocked-out alleles as a donor source for improving other varieties through backcrossing.[16] The resultant soybean varieties with high oleic oil are marketed by Calyxt, Inc. (a subsidiary of Collectis SA) under Calyno™ trademark and monopolistic patent rights for protection of invention.

The fact that biotechnological companies seek patenting of transgenic events or edited traits implies that they give much more importance to traits than totality of characteristics in plant varieties. The number of countries where plant varieties are patentable is limited to the US, Australia, and Japan. Nonetheless, in other countries where there is an explicit ban on patenting of plant varieties, patenting of modified genes is permitted. A recent study on patent legislations of global south reveals patent-eligibility of plant parts (cells, genes, traits, etc.) in 60% of 126 countries

[15] https://www.upov.int/meetings/en/details.jsp?meeting_id=50787.

[16] US Patent 10,113,162.

studied (Oxfam, 2018, 6). This shows that breeders can benefit from extensibility of patented entity to whole plants harbouring it. However, second breeders may inadvertently infringe on patents, as generally no information is given in patent's descriptive claims concerning target varieties. This challenge is addressed by 'PINTO' database that links patents to names of plant varieties.[17] It maintains records of 786 varieties (as of January 2020), albeit incomplete owing to voluntary contribution of patent owners and licensees. Yet defined scope of patented traits, breeders are perplexed by patent thickets, wherein deciphering overlapping set of patent rights on one entity (e.g. trait) is not easily achievable.

Any revision of essential derivation concept, if any, can take advantage of the following facts: development of EDVs is independent of breeding method deployed by breeders, whilst it becomes irrelevant once be generalised to all plant species. Instead, it may be confined to certain species where distinctive morphological characteristics (like petal colour) are of a marketing value. In adopting Australian approach, an emerging challenge would be definition of 'cosmetic change', so as interests of both parties in varietal disputes are kept. The revision should revive breeder's exemption principle for the benefit of society. In this way, implications of patented entities harboured by protected varieties should not be ignored in trait breeding era, where breeding companies prioritise traits within their inventory of assets.

## References


Abe, F., Haque, E., Hisano, H., et al. (2019). Genome-edited triple-recessive mutation alters seed dormancy in wheat. *Cell Reports, 28*(5), 1362–1369.

Annicchiarico, P., Nazzicari, N., Ananta, A., et al. (2016). Assessment of cultivar distinctness in alfalfa: A comparison of genotyping-by-sequencing, simple-sequence repeat marker, and morphophysiological observations. *The Plant Genome, 9*(2), 1–12.

Arens, P., Mansilla, C., Deinum, D., et al. (2010). Development and evaluation of robust molecular markers linked to disease resistance in tomato for distinctness, uniformity and stability testing. *Theoretical and Applied Genetics, 120*, 655–664.

Arora, S., Steuernagel, B., Gaurav, K., et al. (2019). Resistance gene cloning from a wild crop relative by sequence capture and association genetics. *Nature Biotechnology, 37*, 139–143.

ASTA, American Seed Trade Association. (2004). *ASTA position statement on intellectual property rights for the seed industry*. https://www.betterseed.org/the-issues/intellectual-property/asta-position-statement/. Accessed 1 Jan 2020.

Chen, K., Wang, Y., Zhang, R., et al. (2019). CRISPR/Cas genome editing and precision plant breeding in agriculture. *Annual Review of Plant Biology, 70*, 667–697.

Chen, L., & Liu, Y. G. (2014). Male sterility and fertility restoration in crops. *Annual Review of Plant Biology, 65*, 579–606.

Cockram, J., Jones, H., Norris, C., et al. (2012). Evaluation of diagnostic molecular markers for DUS phenotypic assessment in the cereal crop, barley (*Hordeum vulgare* ssp. *vulgare* L.). *Theoretical and Applied Genetics, 125*, 1735–1749.


[17] http://pinto.euroseeds.eu/.

Cockram, J., Norris, C., & O'Sullivan, D. M. (2009). PCR-based markers diagnostic for spring and winter seasonal growth habit in barley. *Crop Science, 49*, 403–410.
Dirks, R., Van Dun, K., De Snoo, C. B., et al. (2009). Reverse breeding: A novel breeding approach based on engineered meiosis. *Plant Biotechnology Journal, 7*, 837–845.
Fradgley, N., Gardner, K. A., Cockram, J., et al. (2019). A large-scale pedigree resource of wheat reveals evidence for adaptation and selection by breeders. *PLoS Biology, 17*, e3000071.
Hajjar, R., & Hodgkin, T. (2007). The use of wild relatives in crop improvement: A survey of developments over the last 20 years. *Euphytica, 156*(1-2), 1–13.
Haudry, A., Cenci, A., Ravel, C., et al. (2007). Grinding up wheat: A massive loss of nucleotide diversity since domestication. *Molecular Biology and Evolution, 24*(7), 1506–1517.
Heckenberger, M. (2004). *Identification of essentially derived varieties in maize (Zea mays L.) using molecular markers, morphological traits, and heterosis*. Ph.D. thesis, University of Hohenheim, https://opus.uni-hohenheim.de/hoptest/volltexte/2004/64/pdf/Komplettversion_3.pdf#page=40. Accessed 1 Jan 2020.
Hills, M. J., Hall, L., Arnison, P. G., et al. (2007). Genetic use restriction technologies (GURTs): Strategies to impede transgene movement. *Trends in Plant Science, 12*, 177–183.
ISF, International Seed Federation. (2012). *ISF view on intellectual property*. https://www.worldseed.org/wp-content/uploads/2015/10/View_on_Intellectual_Property_2012.pdf. Accessed 1 Jan 2020.
ISF, International Seed Federation. (2014). *Guidelines for the handling of a dispute on essential derivation in maize lines*. https://www.worldseed.org/wp-content/uploads/2015/10/ISF_Guidelines_Disputes_EDV_Maize_2014.pdf. Accessed 1 Jan 2020.
Jamali, S. H., Cockram, J., & Hickey, L. T. (2019). Insights into deployment of DNA markers in plant variety protection and registration. *Theoretical and Applied Genetics, 132*(7), 1911–1929.
Janis, M. D., & Smith, S. (2007). Technological change and the design of plant variety protection regimes. *Chicago-Kent Law Review, 82*(3), 1557–1615.
Lesser, W., & Mutschler, M. (2004). Balancing investment incentives and social benefits when protecting plant varieties. *Crop Science, 44*, 1113–1120.
Lombardo, L. (2014). Genetic use restriction technologies: A review. *Plant Biotech Journal, 12*, 995–1005.
Nair, S. K., Wang, N., Turuspekov, Y., et al. (2010). Cleistogamous flowering in barley arises from the suppression of microRNA-guided HvAP2 mRNA cleavage. *Proceedings of the National Academy of Sciences of the United States of America, 107*(1), 490–495.
Noli, E., Teriaca, M. S., & Conti, S. (2013). Criteria for the definition of similarity thresholds for identifying essentially derived varieties. *Plant Breeding, 132*(6), 525–531.
Oxfam. (2018). *The status of patenting plants in the Global South*. The Hague, The Netherlands: Oxfam Novib.
Prifti, V. (2017). The breeder's exception to patent rights as a new type of research exception. *Rights and Science*, 109–116. https://dialnet.unirioja.es/descarga/articulo/6178810.pdf
Sanderson, J. (2017). *Plants, people and practices: The nature and history of the UPOV convention*. Cambridge, UK: Cambridge University Press.
Schinkel, H., & Schillberg, S. (2016). Genome editing: Intellectual property and product development in plant biotechnology. *Plant Cell Reports, 35*(7), 1487–1491.
Smith, S., Bubeck, D., Nelson, B., et al. (2015). Genetic diversity and modern plant breeding. In M. Ahuja & S. Jain (Eds.), *Genetic diversity and erosion in plants* (Sustainable Development and Biodiversity, Vol 7). Cham, Switzerland: Springer.
Tanksley, S. D., & McCouch, S. R. (1997). Seed banks and molecular maps: Unlocking genetic potential from the wild. *Science, 277*(5329), 1063–1066.
UPOV, International Union for the Protection of New Varieties of Plants. (2013). *Guidance on the use of biochemical and molecular markers in the examination of distinctness, uniformity and stability (DUS). TGP/15*. https://www.upov.int/edocs/tgpdocs/en/tgp_15.pdf. Accessed 1 Jan 2020.

UPOV, International Union for the Protection of New Varieties of Plants. (2014). *The use of molecular markers (SNP) for maize DUS testing*. BMT/14/10. https://www.upov.int/edocs/mdocs/upov/en/bmt_14/bmt_14_10.pdf. Accessed 1 Jan 2020.

UPOV, International Union for the Protection of New Varieties of Plants. (2017). *Explanatory notes on essential derived varieties under the 1991 Act of the UPOV convention*. UPOV/EXN/EDV/2. https://www.upov.int/edocs/expndocs/en/upov_exn_edv.pdf. Accessed 1 Jan 2020.

UPOV, International Union for the Protection of New Varieties of Plants. (2018). *Use of SNP markers for soybean variety protection purposes in Argentina*. https://www.upov.int/edocs/mdocs/upov/en/bmt_17/bmt_17_22.pdf. Accessed 1 Jan 2020.

Van Eeuwijk, F., & Law, J. (2004). Statistical aspects of essential derivation, with illustrations based on lettuce and barley. *Euphytica, 137*, 129–137.

Van Wijk, A., & Louwaars, N. (2014). *Framework for the introduction of plant Breeder's rights: Guidance for practical implementation*. Roelofarendsveen, The Netherlands: Naktuinbouw.

Wang, C., Liu, Q., Shen, Y., et al. (2019). Clonal seeds from hybrid rice by simultaneous genome engineering of meiosis and fertilization genes. *Nature Biotechnology, 37*(3), 283–287.

Watson, A., Ghosh, S., Williams, M. J., et al. (2018). Speed breeding is a powerful tool to accelerate crop research and breeding. *Nature Plants, 4*, 23–29.

Zhang, R., Liu, J., Chai, Z., et al. (2019). Generation of herbicide tolerance traits and a new selectable marker in wheat using base editing. *Nature Plants, 5*, 480–485.

# Chapter 7
# Farmers as Researchers: Government Regulation of Farmers' Local Knowledge in Indonesia

**Christoph Antons, Yunita T. Winarto, Adlinanur F. Prihandiani, and Sinta Uli**

**Abstract** This chapter examines the international and national legal frameworks that have fostered the current situation with regards to seeds and local agricultural knowledge in Indonesia. Case studies show how such frameworks are impacting on the practices of farmers and how farmers are dealing with this impact and are developing self-help-mechanisms. It will be argued that standardised regulatory approaches ignore the considerable contribution and innovative nature of local farming practices, which still contribute about 70% of the food produced in the world. The penultimate section of the chapter discusses the changes that can be expected from the new Draft Law on the Sustainable Agricultural Cultivation System. The chapter concludes by pointing to the important influence of two key aspects of the political reformation process in Indonesian society after the end of the military backed Suharto government: decentralization policies which have been beneficial for local councils and NGOs working on rural issues and the creation of a Constitutional Court, which has issued progressive decisions in cases concerning indigenous peoples and farmers. However, while proposed law reform follows the lead of the judiciary, it also reclaims some of the discretionary powers for the bureaucracy and creates new burdens for farmers.

**Keywords** Local knowledge · Agriculture · Seeds · Decentralization · Constitutional Court

This work was supported by the Australian Research Council under the Discovery Projects funding scheme (Project Number DP170100747).

C. Antons (✉)
Faculty of Business and Law, Newcastle Law School, University of Newcastle, Newcastle, NSW, Australia
e-mail: christoph.antons@newcastle.edu.au

Y. T. Winarto · A. F. Prihandiani · S. Uli
Faculty of Social and Political Sciences, Universitas Indonesia, Depok, Indonesia



M. Blakeney, K. H. M. Siddique (eds.), *Local Knowledge, Intellectual Property and Agricultural Innovation*, https://doi.org/10.1007/978-981-15-4611-2_7

## 7.1 Introduction: The Global Standardisation of Farming Practices and Its Impact on Farmers' Freedom to Experiment

In the thinking of classical development economists, industrialization and agricultural transformation go hand in hand.[1] With modernization and technical progress in agriculture, its share of domestic output and of the labour force declines (Nafziger, 2006, 221–222). Accordingly, while agriculture still contributes between 20% and 60% to the GDP of many developing countries, it represents only between 1% and 2% in the EU, United States and Australia (Antons, 2016, 391). Similarly, economists point to an inverse relationship between a nation's per capita income and the size of its rural population (Cypher & Dietz, 1997, 331). Hence, economic development processes are designed to follow the examples of the industrialised countries and to encourage the migration of rural populations to the cities, and to radically transform agricultural technologies and practices, so that they become more efficient and can feed a larger population that works in the manufacturing or service sector and is no longer producing food. In this development model based on the experience of the industrialised countries, agricultural evolution moves from peasant farming over mixed farming to commercial farming, which is understood as the most developed and sophisticated form of farming (Nafziger, 2006, 226–227). New seed varieties and other inputs such as fertilizers are part of this process.

With regard to seed varieties, the Green Revolution of the 1960s introduced new high yielding varieties in Asian developing countries. With the new seeds came seed certification laws that were designed to promote scientifically developed and tested varieties over traditional and local ones (Utomo, 2013, 4). The effects of the Green Revolution have been controversial ever since. Critics point to the inequality and monocultures that were created, with the overuse of pesticides[2] making it now necessary to find solutions against increasingly invasive pests within very short time frames (Fox, 2014). On the other hand, proponents of the Green Revolution contend that food supply would not have kept up with population growth in countries like India without high yielding varieties and technological input. They argue that a new Green Revolution is needed that includes biotechnological applications.[3]

Many governments in developing countries are torn between the need to defend the interests of a still large traditional farming sector and what they regard as the need to catch up technologically with the advanced industrial nations. In law and policies, they often try to promote both aims at the same time. Hence, both India and Indonesia now have laws for the protection of new varieties of plants that promote commercial plant breeding, as well as farmers' rights (in India)[4] and local varieties

---

[1] Lewis, 1954, 119; as quoted in Nafziger, 2006, 221.

[2] See the sources in Nafziger, 2006, 288–290.

[3] Nafziger, 2006, 287–288, with further sources.

[4] See the Protection of Plant Variety and Farmers Right Act, 2001 (PPVFR Act).

(in Indonesia).[5] Their legislative initiatives are constrained, however, by obligations from international treaties related to trade and the environment. As the following sections will show, the shift from traditional farming practices involving local exchange and experimentation with seeds, to commercialised, standardised and scientific practices of plant breeding, has been accompanied by an increasingly restrictive regulatory environment that constrains the initiative of farmers to creatively develop and adapt plant material to local conditions. In the worst cases, this new regulatory environment punishes creativity and criminalizes small scale farmers who experiment with seeds. It will be argued that such standardised approaches ignore the considerable contribution and innovative nature of local farming practices, which still contribute about 70% of the food produced in the world (McKeon, 2015, 3). Agro-meteorologists have explained the importance of the local knowledge of farmers for climate change adaptation in view of the increasing difficulties in predicting seasonal rainfall patterns (Stigter, 2016a, 2016b). Others have pointed to the superiority of local knowledge after natural calamities and catastrophic events (Coelho, 2007).[6] In general, it seems problematic to term forms of knowledge "traditional" (Silitoe, 2006) that are essentially an "admixture of local folk knowledge and extra-local scientific knowledge" (Dove, 2000, 215) that could be termed "peasant science" (Frossard, 2005; Winarto, 2004) and to juxtapose "traditional" and "scientific" forms of knowledge (Agrawal, 1995). Intellectual property laws and other state laws, however, continue to be based on hierarchies of knowledge that have their origins in colonial times and that disregard much of local knowledge as unscientific and superstitious (de Sousa Santos, Nunes, & Meneses, 2007). The following section of this chapter examines the international legal framework that has fostered the current situation. This is followed by an examination of the national framework for local farming practices in Indonesia, and sections with case studies that show how this is impacting on the practices of farmers and how farmers are dealing with this impact and are developing self-help mechanisms. The penultimate section of the chapter discusses the changes that can be expected from the new law on the Sustainable Agricultural Cultivation System (*Undang-Undang Sistem Budidaya Pertanian Berkelanjutan*), which replaced the plant cultivation law in 2019. The article concludes by pointing to the important influence of two key aspects of the political reformation (*reformasi*) process in Indonesian society after the end of the military backed Suharto government: decentralization policies which have been beneficial for local councils and NGOs working on rural issues and the creation of a Constitutional Court, which has issued progressive decisions in cases concerning indigenous peoples and farmers. It will be shown that the law reform follows the lead of the judiciary, but also reclaims some of the discretionary powers of the bureaucracy.

---

[5] See the Law on the Protection of Plant Varieties of Indonesia, in particular section 7 on local varieties.

[6] See also the contributions in Ellen (2007).

## 7.2 The International Framework: The Development of Seed Laws and Intellectual Property in Plant Material

Laws meant to guarantee the quality of seeds and to certify standards appeared in developing Asia over a few decades beginning in the 1960s and coinciding with the Green Revolution. Originally meant to protect farmers from bad seeds being distributed by industry (GRAIN & Sharma, 2005), they still had a somewhat limited role at a time when seed production was dominated by public sector institutions. In India, the Seeds Act of 1966 established a Central Seeds Committee and central and state seed laboratories. The government began to notify regulated varieties and to specify minimum levels of germination and purity as well as labels to confirm these standards. Many other countries in developing Asia followed between the 1970s and early 2000s. By that time, policies encouraging private enterprise in sectors such as agriculture meant that the role of the public sector was much reduced, and seed laws were aimed as much at the encouragement and further development of the seed industry as at the protection of farmers as consumers of seeds (GRAIN & Sharma, 2005, 26).

In that sense, they began to show some similarities with intellectual property laws for plant material, being designed to create incentives for the development of new varieties of plants, and introduced following the conclusion of the 1994 WTO TRIPS Agreement. This Agreement required, in Article 27.3(b), that WTO member states protect plant varieties either by patents or by an effective *sui generis* system, or by any combination thereof. Providing intellectual property rights for plant material was new to the great majority of developing countries at the time. National Patent Acts prior to the TRIPS Agreement had regularly excluded the granting of intellectual property to forms of life (UNCTAD-ICTSD, 2005, 390). Developing countries were now facing a deadline in 2000 to introduce either specialised plant variety laws or patent protection. Most opted for plant variety laws, which offer more freedom to experiment for farmers than patent laws. Because there is an International Convention for the Protection of New Varieties of Plants, known under its French acronym as UPOV, most countries modelled their laws after one of two versions of this convention, one dating from 1978 and the other from 1991 (Antons, 2016).

That developing countries opted for UPOV conforming laws is in itself remarkable, because TRIPS requires systems to be "effective", but does not otherwise prescribe what a national system must look like. UPOV was founded in the 1960s by a group of European countries (Janis, Jervis, & Peet, 2014, 69–71). With its requirement for varieties to be "new, distinct, uniform and stable" it reflects the advanced commercialized agricultural sectors of such countries and these requirements of course preclude the protection of traditional varieties, which are not new, distinct, uniform and stable. Some countries, such as India, decided therefore to introduce elements for the protection of the traditional farming sector into their national laws (Antons, 2007, 2010). This is clearly reflected in the title of the Indian legislation: the "Protection of Plant Varieties and Farmers' Rights Act". Although the private law origins and aims of the intellectual property laws are quite distinct from the consumer protection and public law aims and origins of the seed laws, both are commonly administered by agricultural ministries. As the test planting of the varieties is time consuming and expensive, the conditions for protection of both seed and intellectual property laws are also often tested together, if registration under both systems has been required.

The latest developments at the international level are to be found in multilateral and bilateral Free Trade Agreements. Such FTAs often have intellectual property chapters and they frequently include obligations for developing countries to introduce even higher standards than those established by the TRIPS Agreement. A number of developing countries have promised to introduce patent protection for plants, or to 'endeavour' to do so, while others, including Indonesia, have promised to join the UPOV Convention or to introduce plant variety laws in accordance with the UPOV Convention (Antons, 2016, 2019).

## 7.3 The Domestic Legal Framework in Indonesia Until 2019: The Law on the Plant Cultivation System and the Law for the Protection of Varieties of Plants

When the Law on the Plant Cultivation System[7] was introduced in Indonesia in 1992, it replaced a very short law from 1961[8] that was only concerned with the import and export of plants and seeds, as well as a number of colonial era ordinances for specific areas of agriculture.[9] Enacted during the final years of the strongly development oriented "New Order" of President Suharto, Law No. 2 of 1992 stressed government "guidance" of farmers, as a number of NGOs acting on behalf of farmers pointed out during their constitutional challenge to several provisions of the Act (Mahkamah Konstitusi, 2013a, 24). The government explanation to the Preamble of the Law pointed out that increased efficiency in the agricultural sector was necessary because of a rising population and the fact that much agricultural land had been diverted to other uses. To achieve increased efficiency, the government developed a plant cultivation development plan in accordance with the national development plan, determining the areas of plant cultivation development, and regulating the production of certain cultivated plants in accordance with the national interest.[10]

Much has changed in Indonesia since this law was introduced and, hence, the law was completely revised in 2019. After the criminal provisions of the 1992 law had

---

[7] *Undang-Undang Nomor 12 Tahun 1992 Tentang Sistem Budidaya Tanaman.*

[8] *Undang-Undang Nomor 2 Tahun 1961 Tentang Pengeluaran dan Pemasukan Tanaman dan Bibit Tanaman.*

[9] Article 65 UU No 12/1992.

[10] Article 5 UU No 12/1992.

been used to prosecute several small farmers in the Kediri and Nganjuk regencies of East Java for unauthorised seed certification (Jhamtani & Patria, 2006; Indonesian Human Rights Committee for Social Justice, 2013; Tempo, 2005), a number of NGOs acting on behalf of farmers questioned the constitutionality of key provisions of the law, including those that had been used in the prosecution of the farmers. They achieved a partial victory in Indonesia's Constitutional Court. In one of the most widely publicised cases concerning the farmer Tukirin, the prosecution had relied on Article 14(1) in conjunction with Article 61(1)(b) of Law No. 12 of 1992 on the Plant Cultivation System (Jhamtani & Patria, 2006; Indonesian Human Rights Committee for Social Justice, 2013). Article 14(1) explained that the certification of seeds is the prerogative of the government and of authorised individuals and legal persons. Article 61(1)(b) provided penalties for those who carry out certification without authorisation. Jhamtani and Patria (2006) point out that this was a strange basis for the prosecution in this case, because Mr. Tukirin in fact did not certify anything. He had been breeding his own seeds after having earlier been given seeds to plant in the context of a cooperation between a commercial seed company and the local government. He had no knowledge of the purpose of the project and no contract was ever concluded with him. Still, the judge imposed a one-year probationary jail term and prohibited Tukirin from planting his own seeds for the same period (Jhamtani & Patria, 2006, 2–3).

The arrests and/or prosecutions of other farmers were on the grounds of trade mark violations, of carrying on the business of trading in seeds without authorisation,[11] of searching for and collecting germplasm without approval,[12] of releasing not yet certified varieties[13] or of aiding and abetting in such activities.[14] Farmers subsequently began to organise themselves with the help of several NGOs and they achieved strong press coverage of their cases (Utomo, 2013, 54–55). A few of them joined the NGOs led by the Indonesian Human Rights Committee for Social Justice (ICHS) in their Constitutional Court case reviewing key provisions of the Law on the Plant Cultivation System of 1992.[15] The court agreed with the appellants that Article 9(3) and Article 12(1) of the Law of 1992 were unconstitutional in so far as they made the search and collection of germplasm dependent on permits, and the

[11] Articles 61(1) d., 48(1) of the Plant Cultivation System Law of 1992.

[12] Articles 60(1) a, 9(3).

[13] Article 60(1) b., 12(2).

[14] Article 56 Criminal Code; Utomo (2013), 61–62.

[15] The other NGOs acting on behalf of the farmers were the Farmer Initiative for Ecological Livelihoods and Democracy (FIELD), the Alliance of Indonesian Farmers (*Aliansi Petani Indonesia*), the Sadajiwa Village Cultivation Foundation (*Yayasan Bina Desa Sadajiwa*), the People's Coalition for Food Sovereignty (*Koalisi Rakyat untuk Kedaulatan Pangan*), the Indonesian Farmers' Society for Integrated Pest Control (*Ikatan Petani Pengendalian Hama Terpadu Indonesia*), the Alliance of Oil Palm Farmers (*Ikatan Petani Kelapa Sawit*), the Coconut Watch Association (*Perkumpulan Sawit Watch*), the Union of Indonesian Farmers (*Serikat Petani Indonesia*) and the Alliance for the Agrarian Reform Movement (*Aliansi Gerakan Reforma Agraria*). Individual farmers involved came from the Kediri and Indramayu districts in East and West Java respectively.

release of varieties the prerogative of the government, in both cases without making exemptions for small scale farmers (Mahkamah Konstitusi, 2013a, 124–125). The Court decided that the provisions had to be amended so that they exempted small farmers in the case of Article 9(3), and the release of varieties by small farmers in the case of Article 12(1) (Mahkamah Konstitusi, 2013a, 128–129). A change to the criminal provisions of Articles 60(1) a. and b. and Article 60(2) a. and b., whose constitutionality had also been questioned by the appellants, was no longer necessary as they merely referred back to the articles that the court had reinterpreted and amended in accordance with the Constitution (Mahkamah Konstitusi, 2013a, 126–127). The Court did not elaborate precisely on which articles of the constitution the decision was based. In questioning the constitutionality of the provisions, the appellants had relied on many articles from the human rights chapter XA of the Constitution,[16] which had been introduced with the second amendment of the Constitution in 2000, and on two subsections of Article 33 of Chapter XIV on the national economy and social safety, which stated, among other things, that the natural wealth of Indonesia is controlled by the state but must be used for the greatest possible prosperity of the people.

The Constitutional Court decision was celebrated by commentating experts in the Indonesian press as a victory for the farmers (Santosa, 2013). Undoubtedly, it was a very important decision. It started the process of removing some of the immediate legal reasons for the criminalization of farmers, and triggered the revision of Law No 12 of 1992. However, it was only a partial victory, because the Court disagreed with the appellants on the potentially negative implications of the rural development planning policies of the government and their impact on the independence, creativity and freedom of farmers to plan for themselves. The applicants had also questioned the constitutionality of Article 5 of Law No. 12 of 1992, which was part of the Chapter on the "Planning of Plant Cultivation". According to subsection d. the government created the conditions that triggered the participation of society in the planning processes, which included, as mentioned previously, plant cultivation planning in accordance with the national development plan, determination of the areas of plant cultivation development, and the regulation of plant cultivation production in the national interest.[17] The applicants alleged that Article 5(1) violated Article 28A, C, D, F, I and H as well as Article 33 (2), (3) of the Constitution guaranteeing, among others, the right to earn a living, the right to legal certainty and equality before the law, the right to communication and the receipt of information, the rights of traditional societies and their cultural identity and the right to private property (Mahkamah Konstitusi, 2013a, 22–26). The court disagreed, pointing out that it was the government's responsibility to secure the prosperity of the Indonesian people by planning for development in general and for plant development in

---

[16] Reference was made to Articles 28A, 28C, 28D(1), 28I(2) and (3) and 28G(1), see *Mahkamah Konstitusi* (2013a), 27–37.

[17] Article 5(1) (a), (b), (c).

particular. The various planning responsibilities in this case were in the national interest and in accordance with the Constitution.

The Court also declined to find Article 6 unconstitutional. This article gave farmers the freedom to select the varieties they want to plant and how to cultivate them. In exercising this freedom, however, farmers needed to participate in creating the development and production of plant cultivation as outlined in Article 5. In the situation where government policies prevented farmers from exercising their freedom, the Government was responsible for seeking a guarantee of a certain yield for the farmers. According to subsection 4 of Article 6, further details of this guarantee were to come from a Government Regulation, but this regulation was never issued. The court pointed out that the freedom guaranteed by Article 6 was not unconditional, but limited by the responsibilities of the farmers to participate in the development plan for plant cultivation. According to the Court, even the human rights provisions of the Constitution can be limited by the recognition of and respect for the rights and freedoms of others and by justified demands in accordance with moral considerations, religious values, and public security and order in a democratic society.

The wide discretionary planning powers of the government and the absence of the government regulation promised in Article 6(4) left a considerable amount of uncertainty for farmers, even after the court revised the most immediately concerning provisions of Law no. 12 of 1992. Importantly, also, the Constitutional Court can only revise official laws (*Undang Undang*), but not implementing regulations of the government (*Peraturan Pemerintah*).[18] Such implementing regulations are not automatically revised or invalid when a new law comes into force, unless this is expressly declared in the new law (Antons, 2009, 101–102). The new Law on the Sustainable Agricultural Cultivation System of 2019, discussed in detail below, in Article 124 leaves all implementing laws and regulations issued for Law No. 12 of 1992 in force as long as they are not in contradiction with the Draft Law. For the matters covered in the Constitutional Court decision, Government Regulation No. 44 of 1995 on plant nursery is the main implementing decree that remains relevant. It again refers on many occasions to further decrees at Ministerial level, which also all remain relevant.

Government Regulation No. 44 of 1995 on Plant Nursery repeats many of the provisions from Law No. 12 of 1992 on the Plant Cultivation System that the Constitutional Court regarded as problematic. For example, Article 3(1) explains that germplasm is controlled by the state and has to be used for the greatest possible welfare of the people. Article 5 points out that the Government is in charge of searching, collecting, using and conserving germplasm and that this can be carried out by Indonesian individuals or legal entities, but only with permission from the Minister. The details of the permit process can be collected from the Regulation of the Minister of Agriculture No. 67/Permentan/OT.140/12/2006. Article 6(2) of

[18] Articles 1 No. 3 and 10(1) of the Law on the Constitutional Court (Law No. 24 of 2003 as revised by Law No. 8 of 2011); see also Butt & Lindsey, 2009, 274–275.

Government Regulation No. 44 of 1995 adds that a Ministerial official must accompany any search or collection of germplasm. For the release and circulation of seeds, however, the Government Regulation is narrower than Law No. 12/1992 and mainly concerned with "improved seeds" (*benih bina*), which according to Article 13(1) Law No. 12/1992 are seeds derived from a variety that has been declared after testing as a "superior variety" (*varietas unggul*). The Regulation declares that seeds from superior varieties can only be circulated after they have been released by the Ministry,[19] usually the Ministry of Agriculture, but possibly also the Ministry of Forestry for varieties from forested areas (see the government explanation to Article 1 No. 8) and that the production of 'improved seeds' of a certain scale requires the approval of the Ministry. The approval process is regulated in the Regulation of the Minister of Agriculture No. 56/Permentan/PK.110/11/2015 Regarding the Production, Certification and Circulation of Improved Seeds of Food Crops and Green Fodder for Livestock. This seems to leave more scope for local varieties and farmer varieties that are neither recognised as "superior" or "improved varieties" in the sense of Law No. 12/1992, nor produced on a more extensive scale. However, while Government Regulation No. 44/1995 in this sense does not immediately conflict with the Constitutional Court's more liberal interpretation of Law No. 12/1992, the Regulation as well as the implementing Ministerial Regulation No. 67/2006 on Conservation and Use of Plant Genetic Resources[20] do conflict with the decision and the new law of 2019 in that they do not foresee the exemption for small farmers in collecting germplasm.

The further implementing Regulation of the Minister of Agriculture No. 61/Permentan/OT.140/10/2011 Regarding the Trial, Evaluation, Release and Collection of Varieties was similarly focused on "superior varieties". A list of documentation requirements and of adaptation and observation tests were, as the following case studies and experiences in the field demonstrate, difficult to comply with for the majority of farmers and, as far as they were focused on varieties being "distinct, uniform and stable"[21] unsuitable for traditional or local varieties. "Local varieties" were defined in accordance with the Plant Variety Protection Act and could only be released if they had been registered with the Centre for the Protection of Plant Varieties and Authorisation of Farmers. As a result of the Constitutional Court decision exempting small farmers from some of the requirements of Law No. 12 of 1992, this Regulation was revised in 2017. The new Regulation of the Minister of Agriculture No. 40/Permentan/TP.010/11/2017 Regarding the Release of Plant Varieties will be discussed below.

Law No. 29 of 2000 Regarding the Protection of Plant Varieties was introduced to meet the requirements and the deadline of the TRIPS Agreement, discussed above, to adopt either patent protection or *sui generis* protection for plant varieties.

---

[19] Article 21 Government Regulation No.44/1995.

[20] *Peraturan Menteri Pertanian No. 67/Permentan/OT.140/12/2006.*

[21] Article 13(3) (d).

As did many other developing countries,[22] Indonesia opted for largely UPOV conforming plant variety protection as an internationally acceptable, but less stringent option for protection, in comparison with patent protection (UNCTAD-ICTSD, 395), following with few exceptions the 1991 version of the Convention. Compliance with UPOV 1991 has meanwhile also been promised in the Economic Partnership Agreement with Japan and in the Comprehensive Economic Partnership Agreement with the EFTA countries (Antons, 2019, 248–249).[23] Like India, Indonesia has so far attempted to provide not just incentives for plant breeders to breed new varieties, but to provide in the same law also for Indonesia's agricultural heritage in the form of local varieties.[24] However, while innovative varieties, which must be new, distinctive, uniform and stable,[25] are included in Chapter II, Part One of the Law as "Plant Varieties which can be granted protection", local varieties are only mentioned in Part Five of the same chapter under the heading "Rights and obligations of the right holder of plant variety protection." This raises the question of what is meant by Article 7(1) if it states that local varieties are "owned by the community" but "controlled by the state". Again, Article 7(4) refers to an implementing regulation for the details. This regulation is Government Regulation No. 13 of 2004 Concerning the Naming, Registration and Use of the Initial Variety for the Making of Essentially Derived Varieties. Depending on the geographical spread of the variety, it empowers the governor of a province (*Gubernur*) or District *(Bupati*), the Mayor of a city (*Walikota*) or, where the variety is spread over several provinces, the Plant Variety Registration Office in the Ministry of Agriculture, to represent the community and register the variety on its behalf. Any individual or legal entity using a local variety to produce an essentially derived variety needs to conclude an agreement with those authorities representing the community. Interestingly, the variety owning community is not exempted from this requirement. While some form of compensation for the community 'may' be included in the agreement, such inclusion is not compulsory. If compensation is included, the authorities are required to use it at their discretion for several broadly worded public interest purposes: raising the prosperity of the community; conservation of the local variety; and conservation of genetic resources in the locality of the variety.[26] It has been observed that the regulation leaves the communities without any real rights, but leaves considerable discretion to the authorities whether to negotiate compensation for the communities and how to use such compensation on their behalf (Antons, 2017, 256–257).

As has been pointed out above, there are numerous links between the intellectual property law related to plant varieties and the plant cultivation law with its implementing degrees. The administration of both is the responsibility primarily of the

[22] For the ASEAN Countries, see Kanniah and Antons (2012).

[23] The obligation in the EFTA Agreement is modified by a footnote stating that the relevant provision shall be without prejudice to the rights of Indonesia to protect its local plant varieties (Antons, 2019, 249).

[24] Article 7 of Law No. 29/2000.

[25] Article 2.

[26] Article 10.

Ministry of Agriculture, whereas other intellectual property laws are administered by the Ministry of Justice. Until the revision of the Ministerial Regulation in 2017, release of a local variety could only occur after the local variety had also been registered in the plant variety protection register. Finally, because the certification requirements for the release of a variety under the previous Regulation included that the variety had to be "distinct, uniform and stable" (*unik, seragam dan stabil*), these were partly identical with those of the plant variety protection legislation. As a result, the examination processes for seed certification and plant variety testing could be harmonised and carried out together and at the same time. This possibility remains in the new Regulation of 2017,[27] although it is now made clear that the requirements of novelty, distinctness, uniformity and stability are applied only for the plant variety testing. Still, the intellectual property rules are, therefore, clearly important for farmers, especially where they make use of local varieties and of this combined certification and testing process. In spite of this, however, the following case studies from fieldwork in West, Central and East Java show that so far there is little understanding of them and farmers' discussion so far remains focused on the seed registration requirements.

## 7.4 Struggling with Legal Issues: Farmers' Plant Breeding in the "Global" World

In the discussion about intellectual property rights and conservation, Brush (1996, 1) identifies two types of plant resources and related local knowledge: "…(1) crop germplasm and farmer knowledge of domesticated plants, and (2) natural products derived from wild plants and knowledge about the plants and products." Since the mid of the 2000s, a group of farmers in the regency of Indramayu in West Java and in other regencies in Indonesia have produced another type of knowledge with their products, resulting from the dialectic between local and scientific knowledge in plant-breeding (see Winarto (ed.), 2011). Such knowledge is missing in Brush's categories. A number of rice farmers have become "plant-breeders" and are able to produce their own cultivars in the way scientific plant-breeders breed new varieties. Their methods of producing new cultivars imitate scientific ones, though they do not release their products in the "formal market". The local plant-breeders experience a "transition from their cultural practices of selecting, cultivating and exchanging local and state-released-varieties to producing 'locally new-bred rice strands' through adopting plant-breeding scientific premises and methods" (Winarto, 2011a; Winarto & Ardhianto, 2011a). Accordingly, they have been exposed to legal issues of releasing new varieties in a similar way as the formal system of plant-breeding and distribution institutionalized by the state. Since the middle of the first decade of the 2000s, the plant-breeders in Indramayu and some other regencies in Java have

[27] See Article 13 of Minister of Agriculture Regulation No. 40/2017.

been accused by government authorities and industry of having committed “illegal” acts in terms of the formal legal domain of plant breeding (Winarto, 2011b).

Among current trends connecting local communities, environmental change, and international trade in the debate about extending intellectual property rights, one trend has been the “…increased pressure for nation-states to implement intellectual property protection and to conform to a common international standard” (Brush, 1993, 653). This and similar trends, “…raise questions about the legal status of indigenous groups and their control over culturally specific but widely useful information, and point out the need to conserve both biological resources and indigenous knowledge” (Brush, 1993, 653). Despite the positive objective of intellectual property law as a means of returning economic benefits to local communities for the use of their knowledge and resources (Brush, 1993, 653), this objective is difficult to implement and farmer-plant-breeders have been constrained in their activities for producing their own new rice cultivars.

Farmer-plant-breeders thus have a dilemma to resolve. On the one hand, following the decision of the Constitutional Court discussed above and the revision of the plant cultivation law discussed below, they have greater legal freedom now to pursue their activities. On the other hand, they still have to cope with the formal regulation imposed by the nation-state in dealing with the issues of intellectual property rights and the planning of their activities in the context of development policies. How do the farmers cope with this conflicting situation?

### *7.4.1 Costs and Bureaucratic Requirements of the Certification Process*

The first step the farmer-plant-breeders had to take to release their cultivars—according to the Ministry’s Regulation no. 61/2011 Regarding the Trial, Evaluation, Release and Collection of Varieties—was to have their varieties tested in multiplication or adaptation tests at several limited sites (*Uji Adaptasi* or *Lapangan Uji Terbatas [LUT]*). In this kind of test, the number of examination sites would depend on the kind of crop to be examined. Prior to the test, the farmer-plant-breeders had to submit a written request to the Ministry of Agriculture through the Head of the Research and Development Agency and to complete various forms as part of the administrative requirements. To reach the stage of examination in the adaptation test, several steps had to be taken by the plant-breeders:

- They must have a clear genealogy of the cultivar, so that its genetics are traceable.
- They must have a complete description of the cultivar so that the cultivar can be identified accurately.
- Their cultivar must have prominent features in comparison with other cultivars.
- The cultivar should be unique, uniform, and stable (reaching F8, referring to the planting of one generation (filial) per season until Filial (generation) 8 is reached).

- The plant-breeders had to declare that they had a good quality cultivar and an abundance of seeds to continue multiplying it.

If the cultivars met all the requirements, within 10 days the Research and Development Agency of the Ministry of Agriculture recommended that an adaptation test can be carried out. This adaptation test was carried out at 16 sites for two planting seasons in a row, with detailed examination of the plant's characteristics. Following the adaptation test, the products were evaluated by the Evaluation Team for Releasing Varieties (*Tim Penilaian Pelepasan Varietias*) as a basis for a further decision by the Ministry of Agriculture.

Not only were the procedures lengthy, but the costs were also excessive, amounting to as much as USD42.900. For farmers, of course, such an amount is beyond their ability to pay, and thus it was out of the question for them to follow the official procedures for releasing their cultivars. "Many farmer-plant-breeders decided to stop being plant-breeders because of those enormous costs. Since I like doing breeding, I keep continuing it," said Yus, an Indramayu plant-breeder. If Dar, another Indramayu plant-breeder, was successful in gaining the Ministry of Agriculture's approval for his original cultivar to be released, it was possibly due to the help of the NGO FIELD Alliance, who at the time supported the Indonesian Farmers' Alliance for Integrated Pest Management (IPPHTI) of the Indramayu Regency for that particular purpose.

Even though farmers' rights and ability to be plant-breeders were to some extent accommodated through the Constitutional Court's revision of Law No. 12/1992 on Plant Cultivation Systems (*Sistem Budidaya Tanaman*), the authorities still perceived farmers as incapable of knowing the principles of breeding plants and the genetics of plants. From the viewpoint of the authorities, scientists have exclusive knowledge of this subject-matter (See Cleveland, Soleri, & Smith, 2000; Simmonds, 1979). The now abolished Ministry's Regulation No. 61/2011 and other regulations[28] have been proof of that paradigm.

### *7.4.2 Problems with Labels*

The Ministry of Agriculture's Regulation on the Production, Certification, and Distribution of Improved Seeds of Food Crops and Green Fodder for Livestock of 2015[29] states that seeds are divided into 4 categories based on their functions. The Indonesian government made efforts to sustain the quality of high yielding varieties through various standardization stages as a way to ensure the quality of released

[28] Balai Litbang Pertanian. "Regulasi: Peraturan Mentan No. 61/Permentan/OT.140/10/2011 Tahun 2011" http://www.litbang.pertanian.go.id/regulasi/one/19/ (last accessed December 21, 2018).

[29] *Peraturan Menteri Pertanian Republik Indonesia Nomor 56/Permentan/PK.110/11/2015 Tentang Produksi, Sertifikasi, dan Peredaran Benih Bina Tanaman Pangan dan Tanman Hijauan Pakan Ternak.*

varieties. Before being multiplied and distributed widely to farmers, the released varieties are required to undergo a certification process to obtain 'labels with particular colours' as indications of the type of seeds and their readiness to be distributed. There are four coloured labels for seeds in the certification process, namely: (a) a yellow label for breeder seeds under the control of plant-breeders/institutions as the source for multiplying foundation seeds; (b) a white label for foundation seeds as the first generation of breeder seeds; (c) a violet label to identify seeds as stock seeds, the generation of breeder seeds produced to keep sustaining the identity and originality of seeds to fulfil the quality standard; and (d) a blue label to identify extension seeds as the generation of the foundation seeds which have to be cultivated such that their identity and originality can be sustained. The latter two have to be certified by the Agency of Controlling and Certifying Seeds (*Balai Pengawasan dan Sertifikasi Benih* [BSPB] of the Ministry of Agriculture (Balai Besar Penelitian Tanaman Padi 2016).

Farmers do not have much detailed knowledge of the functions of all the labels and the differences between them. None of the farmer-plant-breeders identified those labelling stages. Only in discussions about farmers' choice of rice varieties in the midst of the ongoing climate variability did some farmer-rainfall-observers raise the issue of the label of a particular variety subsidized by the government, and which did not perform well (*Mekongga*). The variety was not uniform and limited to rice production. The variety was released by the government agency (Indonesian Center for Rice Research) and multiplied by Sang Hyang Sri, the government-owned seeds company. In the farmers' eyes, it should not have delivered such a bad result.

Farmers' ignorance of the differences of labelling was used by those providing subsidized seeds, namely the government officials and shop owners. Seeds which were, in fact, still at the stage of having the white or violet-labels were distributed as subsidized ones, which are supposed to have reached the blue-label level. Farmers could, therefore, incur great losses, if they followed these indicators.

### 7.4.3 Lack of Recognition of Locally Bred Varieties

Without any formal recognition by the authorities in regencies of farmers' cultivars, the farmers' efforts to help their fellow farmers is constrained, because they only have permission to circulate their cultivars within their own local community. The farmers' own cultivars are being categorized as "farmer varieties" and as different from "modern varieties" (See Cleveland et al., 2000). "Modern varieties" have been examined across different areas and ecosystems and thus should have a more uniform and higher quality, with greater productivity than the farmers' variety, which is perceived as suitable only within its own local ecosystem.

The Indramayu regency authority knew of farmer-plant-breeders' activities since 2003, when it encountered efforts by IPPHTI in showing the farmers' capability in plant breeding in the film *Bisa Dèwèk* (We Can Do it Ourselves). Although the regency authority then agreed to provide funding for Farmers' Participatory Plant breeding Schools in 2008, there has been no legal recognition of farmers' own cultivars up to now. There were no follow-up programs by the regency authority to facilitate and improve farmers' plant-breeding. Farmer-plant-breeders have therefore felt disappointed. They are sure that their own cultivars are suitable for the geographical and ecosystem conditions of Indramayu, because they followed the selection process for up to 10 planting seasons. Throughout this long selection process, various factors as well as farmers' own preferences have influenced the end product they themselves prefer (See Winarto & Ardhianto, 2011a, 2011b). As Lyon (1996, 40) says: "[f]armers' evaluation of their own research will be determined by the criteria that they deem most important and such criteria may be intimately connected to local conditions or individual preferences." Accordingly, other farmers could choose which varieties they would like to cultivate based on the characteristics of the cultivars (the form and colour of the grains and husked-rice, age, height of plants, yields, and taste). In the view of farmer-plant-breeders, the potential should be utilized by the regency authority and such local varieties should be developed and promoted as the regency's prominent assets. Unfortunately, this has never been the case up to now.

### 7.4.4 *Criminalizing Farmers*

In view of the aims of uniformity and standardization, farmers were not supposed to produce and release new cultivars. That was the government's response in the early stages of plant-breeding activities by some of the Indramayu farmers (Winarto, 2011).[30] Farmers were not only harassed by government agencies as in Indramayu, but also by seed corporations as experienced by farmers in Kediri, East Java. More recently, criminal charges were laid against a Village Head and Director of a village corporation in Aceh (Kompas, 2019). As described in Part 3 of this article, in the cases in Kediri, a foreign-investment seeds company sued a group of maize-farmers who allegedly had pirated their seeds. One farmer (Kun) was brought to court and jailed for a 10 month period. While the intellectual property related charges were not pursued further, he was jailed for selling uncertified seeds. This event jeopardized Kun's livelihood and he ceased his activities in breeding maize.

> Kun, a farmer from a village in Ngasem, Kediri Regency, was found guilty of illegal plant breeding practices, violating the regulations of Law No. 12/1992 of Article 61 (1) d., Article 48 (1) and Article 12, 14, and 60. He had been charged with 'distributing seeds which have not been released by the government and have not been certified'. Being a landless farmer, Kun had learned how to breed maize using a field owned by another farmer. Once he was able to produce the cultivars, he sold the seeds to his fellow farmers without brand name or label. Initially, nobody complained about this practice. However, he was caught by the police on 16th of January 2010, when they were following up on a seeds company report alleging that Kun had pirated the company's cultivars. When they arrested Kun, the police also took 2.5 tons seeds and his equipment.

Farmers were used to breeding their own maize-cultivars and to exchanging seeds with their fellow farmers. Fellow farmers regarded Kun's seeds as a high quality product, resistant to pests, maturing quickly, easy to peel to obtain the corn seeds, and having a nice colour. The price of Kun's seeds was also more reasonable than that of the released seeds. In fact, a genetic test in the laboratory of the Indonesian Center for Biodiversity and Biotechnology (ICBB) in Bogor found that the farmer's seeds and the corporate seeds were entirely different, so there was no basis for an intellectual property claim.

This case motivated the judicial review in the Constitutional Court. Kun also joined the movement and was able to prove his genuine work with the parental seeds he used in plant breeding. As explained in Part 3 of this article, on 18 July 2013, the Constitutional Court approved the farmers' request related to two articles.[31] Although the court did not address some of the other issues comprised in their claim, the farmers had regained their freedom to carry out their plant breeding activities.

How fragile this freedom still is, however, became clear with a subsequent development in Kun's case. He had been hired by a seed-company to produce new cultivars for that company. The police arrested him again, however, acting on a report from fellow farmers, who themselves were afraid of being caught because of intimidation from the seed companies. It turned out that the Constitutional Court decision had not been widely disseminated and the police and other government agencies were unaware of it. Kun was represented by a pro bono lawyer from Surabaya, who provided the authorities with the Court decision. Kun was then released and officially acknowledged as a plant breeder. He was subsequently recruited to engage in plant-breeding on behalf of the Agricultural Research Agency in Kediri. Indramayu plant breeders have had similar experiences. In spite of the Court decision, the Indramayu Regency Agricultural Office has still been questioning farmers' as to whether their plant breeding practices and products have been officially permitted.

The latest case of criminal charges against farmers occurred in July/August 2019, when Mr. Munirwan, the head of Meunasah Rayeuk village in North Aceh was arrested and charged with selling uncertified rice seeds in violation of the relevant provisions of the 1992 Plant Cultivation System Act. It emerged that Munirwan, a local representative of the Indonesian Farmers' Association for Seed Bank and Technology (AB2TI—*Asosiasi Bank Benih dan Teknologi Tani Indonesia*) had introduced IF8 rice seeds originally obtained from AB2TI in his village and villagers had been able to obtain magnificent yields using these seeds. They subsequently adapted the seeds further to the local conditions and sold them via the village owned enterprise to gain further income, whereby they continued to use the AB2TI IF8 label. The case attracted much media attention, because the variety had won awards for village innovation. While the authorities stressed in this case that the village company had sold the seeds in a commercial manner and Mr. Munirwan was arrested as Director of this company, the Constitutional Court's reinterpretation of Article 12

[31] Article 9 (3) on the activities of seeking and collecting genes, and Article 12 (1) on the release of varieties.

of the 1992 Plant Cultivation System Act with its exception for small farmers was not mentioned or discussed (Antaranews, 2019; Harian Aceh, 2019; Putra, 2019; Republika, 2019).

### *7.4.5 Understanding Intellectual Property Rights*

For generations, farmers have been domesticating their own cultivars by the process they call '*menyeleksi*', which means selecting rice panicles which look similar to their parental seeds. The selected grains were then being cultivated in the next planting season. Those local varieties, including their segregated strands, were named "*padi petani*" (farmers' rice) or "*padi lokal*" (local rice). They produced and exchanged those seeds among themselves by various means. Wherever the seeds were planted, knowledge of the seeds grew. Farmers have never considered the seeds or their knowledge of those seeds as something "private". In their view, both the knowledge and the local seeds are the farmers' common property and, thus, do not need to be protected in a formal legal manner that acknowledges special rights for a particular farmer or a group of farmers. At the time the Indonesian government introduced high yielding varieties (HYVs) in the early 1970s, farmers were forced by village officials, supported by the Army, to plant "*padi pemerintah*" (government paddy) or "*padi pendek*" (dwarf paddy), which was unlike their own paddy. Suharto, Indonesia's President in that period, declared the rice intensification program as state policy in the *Bimbingan Massa* (BIMAS, Mass Guidance) and *Intensifikasi Massa* (INMAS, Mass Intensification) measures to replace traditional varieties with high yielding ones, and to include the package of irrigation, chemical fertilizers and pesticides (Fox, 1991; Winarto, 2004).

When some farmers in the Indramayu regency became plant breeders in the middle of the first decade of the 2000s, a partial change in attitude and in terminology could be observed. "*Benih saya*" (my seeds) or, referring to a particular variety, "*Borang ciptaan saya*" (*Borang* my creation) were new terms identifying a particular cultivar as an individual's own seeds or individual creation, a product of plant breeding activities (Ardhianto, 2011). "Legal rights" for their products became commonly discussed whenever farmer-plant breeders had a chance to meet. The idea of intellectual property was introduced by facilitators, state plant-breeders, NGO personnel, and district agricultural officials in the Participatory Plant-Breeding Farmer Field Schools (PPB-FFS or *Sekolah Lapang[an] Pemuliaan Tanaman [Partisipatoris]*). Although the main protection available to such breeders is plant variety protection, the popular term in farmers' discussions to refer to intellectual property rights for a while was patents (*hak paten*). After being introduced by the director of the Plant Variety Protection Office (PVP or in Indonesian *Perlindungan Varietas Tanaman, PVT*) to plant variety rights, farmer-plant-breeders also came to recognise this type of intellectual property rights. Since then, Indramayu farmer-plant-breeders have differing views on whether there is a need to acquire intellectual property rights or not (Ardhianto, 2011).

The motivation for farmers who thought that going through the official procedures as defined by the PVP office was necessary, was not only for gaining legal recognition, but also for achieving financial benefits in the form of profits for their product through the market, and for protecting their seeds from becoming appropriated by private seed companies. Thus, in 2007, Dar was strongly in favour of getting the "patents" for his cultivars so as to prevent any 'pirating' by the seed companies. On the other hand, Arifin, another plant breeder, initially preferred to just get his cultivars officially recognised at the village level through a "community registration" process (Ardhianto, 2011). In the end, however, this 'community registry based on village regulation (*Peraturan Desa*)' failed to materialize and all varieties had to be processed and released via the central PVP office. Following the Constitutional Court decision confirming the right of farmers to pursue their plant breeding activities, Arifin has felt secure in continuing his plant breeding and disseminating his products (up to 19 cultivars) widely via social media and the Plant-breeding Farmer Field Schools he initiated. He has built up his own network to disseminate the knowledge, skills, and products of plant breeding under the name of *Amarta Padi* (*Asosiasi Masyarakat Tani Padi Indonesia*). He insists, however, that there is a need for a regulation to get a "royalty" for his products if other farmers multiply them. This need was also voiced strongly by Dar in 2007 and 2008 when his cultivars were multiplied and sold by other farmers.

Considering the need to get the state's legal recognition for the farmers' cultivars, H. Roni as the head of the Indonesian Integrated Pest Management Farmers' Alliance of the Indramayu Regency (*IPPHTI Kabupaten Indramayu*) initiated a formal collaboration with the Indonesian Center for Rice Research (ICCR or in Indonesian *Balai Besar Penelitian Tanaman Padi)* in Sukamandi, Subang Regency, in 2013.[32] Under the collaboration, several cultivars produced by four Indramayu farmer-plant-breeders were selected to be processed further for an official release through the seeds' certification process. Based on observation, multi-sites-adaptation tests, and a yields procurement test, only Dar's cultivars (*Pemuda Idaman* and *Gadis Indramayu*) from the breeding of *Ciherang* (one of the state's high yielding varieties) and a local rice strand (*Kebo*) were successful in being selected for further process. Dar's cultivars were considered as meeting the criteria of being resistant to the brown planthopper, a common and very destructive pest, having a strong stem-tissue, and producing higher yields than *Ciherang*. However, without any "nucleo-seeds" provided by Dar for his cultivars, the seeds-agency used the seeds provided for the selection test. Those seeds were selected further for several seasons for the seeds-agency plant-breeders' examination until reaching stability and undergoing multi-sites-adaptation tests. Therefore, in the official certified letter released by the Ministry of Agriculture, Dar was mentioned as only one plant-breeder among

---

[32] In line with the tasks and functions of ICRR in research and dissemination, they offer three kinds of collaboration: research and development, dissemination, commercialization, and public services. For further detailed information of ICRR's public services, see '*Layanan*', at Balai Besar Penelitian Tanaman Padi. http://bbpadi.litbang.pertanian.go.id/index.php (last accessed December 21, 2018).

several other state plant-breeders. Since the cultivars were produced for irrigated rice-fields, the new variety was released under the name of *Inpari Agritan 44*. The number 44 was its position in the order of the varieties released for irrigated rice fields under the name of *Inpari* (*Inbrida padi irigasi*). Dar never received any royalty or compensation, except an award provided by the Ministry of Agriculture. In fact, all varieties released by the state agency under the *Inbrida* name such as *Inpari, Inpago, Inpara* were unable to be the subjects of any royalty.

It is apparent that meeting the state's requirement is not easy for farmers who have just gained their novel knowledge and skills in plant-breeding during the last decade. In an open environment of cultivating paddy without any of the established social-cultural institutions of formal plant-breeding as standardized by the state, farmer-plant-breeders' chances to have their cultivars legally certified by the state are very small.

One required component of that standardized process which farmer-plant-breeders cannot easily fulfil is to provide details of the plant's genealogies with precise and proper documents. The case of the Indramayu plant-breeders reveals that they used to breed a government released high yielding variety, e.g. Ciherang, with their local variety. In the PPB FFS held in Indramayu, the methods of documenting the names of parental seeds in plant-breeding and of the filials were introduced to the farmers. In their own way, each farmer-plant-breeder documented their breeding and selection in diverse forms (Winarto & Ardhianto, 2011b). Without any detailed knowledge of the parental seeds and their selection process, however, it is difficult for farmer-plant-breeders to get certified recognition and official release by the state. Examples for these processes are two cultivars produced by farmer-plant-breeders in the province of Lampung under the names of *Sertani* and *Kabir.* According to the state-plant-breeders, both parental seeds of those varieties originally came from the state's plant-breeding station. In the government's view, those varieties should not be distributed widely without having complied with the state's official certification process.

It appears that the state agencies have not done enough to facilitate the activities of an increasing number of farmer-plant-breeders producing cultivars for their own needs and interests. The gap between farmers' knowledge and practices and the state's formal legal regulation persists.

## 7.5 Striving to Account for Farmers' Own Needs: Self-Help Mechanisms

Despite the above mentioned constraints from the state's standardized official requirements for the certification and release of farmers' cultivars, farmer-plant-breeders have persistently continued their practice of breeding their own rice strands. The farmers that have not been breeding themselves have also been continuously improving their learning about how to select the most resistant varieties in the midst of ongoing environmental vulnerabilities as a consequence of climate change, government policies, and the farmers' own conduct. They are not alone in this struggle. A group of scientists from several universities, several non-profit organizations, and some donor agencies are supporting their efforts. In spite of this support, however, farmers also experience constraints when it comes to continuing their practices.

### 7.5.1 *Farmers' Plant Breeding Under Constraining Circumstances*

Becoming a plant-breeder is a unique experience for some farmers which has not been shared widely by the rest of the farming community. Most farmers have limited land for cultivation. Breeding plants means that they need sufficient space to plant the filials for further selection. Further, their activities need to be supported by detailed documentation. The success of the judicial review process does not guarantee that farmers will gain the support of the authorities and be facilitated in their activities by them. The following case studies will demonstrate their different approaches to these challenges.

#### 7.5.1.1 Getting Old and Relying on External Agency Support

Toto, an elderly plant-breeder from a village in Purbalingga regency in Central Java has produced 3 cultivars and preserved as many as 74 local varieties. Since he is getting old, none of his children are interested in becoming plant breeders, and his rice fields are limited, he has decided to give his cultivars and all local varieties to the Indonesian Farmers' Association for Seed Bank and Technology (*Asosiasi Bank Benih dan Teknologi Tani Indonesia*, abbreviated AB2TI), an association in Bogor set up to help farmers save their own seeds. He is now pursuing his activities as the 'multiplier of seeds' sold by a Catholic Church in Central Java. With the limited size of rice fields that he has, he can only plant seeds in rotation.

#### 7.5.1.2 No Lands Are Available: *In Situ* and *Ex Situ* Preservation

After his success in getting his cultivar released by the Ministry of Agriculture, Dar received official acknowledgment when the authorities of other regencies asked him to produce cultivars for their own needs with particular local characteristics. He was able to produce one cultivar for a regency authority in Sumatra, but failed to get any proper reward from this regency authority. In the end, the regency authority did not send anybody to collect the cultivar produced by Dar despite his hard work in breeding and selecting the cultivar and his high costs in planting the seeds. Since this failure to profit from his work, he decided to cultivate his seeds by using plastic buckets in his own yard. In addition to such *in situ* production and conservation, he also preserves his seeds in the freezer for *ex situ* conservation (Ansori, 2011). From his own observation and experimentation that 15–20% of the seeds preserved in this way over a period of 8 years would still grow, and over a period of one to 2 years more than 50% would do so, he finds it safe to keep his seeds in the freezer.

Since Dar has never received from the government the seeds named *Inpari Agritan 44,* which he initiated and helped to develop, he has never been able to distribute this cultivar among his fellow farmers. He kept producing and distributing his earlier cultivars, *Gadis Indramayu* and *Pemuda Idaman* by using social media (facebook, whatsapp, mobile phones and internet assisted by AB2TI by using their facebook page) or direct transaction with the label of AB2TI.

#### 7.5.1.3 Collaborating with a Political Party in Selling Own Seeds

Arifin, who kept continuing his plant-breeding activities prior to and in the aftermath of the making of the film *Bisa Dèwèk* in 2006–2007,[33] has produced around 19 cultivars with complete and systematic documentation. Since his cultivars reached F11 (Filial 11) he decided not to document the characteristics any longer. Although he also participated in the selection competition during collaboration with the government, he received no explanation as to why his cultivars were not successful and were not selected for release by the agricultural research station. He subsequently decided to sell his own cultivars directly. "The procedures to get the cultivars released by the government are too complicated," argued Arifin. Thus, to distribute his seeds, he decided to collaborate with a political party using the label of *Amarta Padi* for them.

Following the Constitutional Court's decision, he regards his activities of selling his own cultivars as legal. He believes that he can also distribute the seeds elsewhere in Indonesia without any boundaries, but he acknowledges on the label that the seeds are for farmers' communities. For Arifin like for Dar, social media and direct transactions are the means to promote and sell his seeds.

#### 7.5.1.4 Distributing Seeds Through Rainfall-Observers' Associations and Exhibitions

Yus, another Indramayu plant-breeder, who persistently cultivates organic rice, decided to keep selecting the various segregations of his cultivar (*Bongi*). Unlike Arifin, he does not bother with detailed documentation of the segregation and

[33] Winarto and Ardhianto (2011a); Ardhianto (2011); Ansori (2011). Arifin was one of several Indramayu plant-breeders who were prominently featured in the film. He was portrayed as a young skilful plant-breeder who had his own group of plant-breeders in his village. In comparison to other plant-breeders, he and Dar persistently continued breeding activities up to recent times (in 2018).

selection process of his cultivar. He develops his knowledge of the characteristics of the seeds by carrying out farmer-field-experiments provided for in the Science Field Shops (SFSs). The Science Field Shops are a forum for learning agrometeorology together with other rainfall-observers facilitated by the Center for Anthropological Studies of Universitas Indonesia (Winarto, Stigter, & Wicaksono, 2017; Winarto, Walker, Ariefiansyah, Prihandiani, Taqiuddin, & Nugroho, 2018). To distribute the seeds, he uses his own farmer-to-farmer networks, the rainfall-observers' association, and any exhibition event in various kinds of meetings within and outside the Indramayu Regency.

### 7.5.2 Finding Alternative Ways for Seed Conservation, Certification, and Distribution

After gaining experience and learning the lessons from the initiatives carried out by IPPHTI in getting official release by the government—which only Dar's cultivar (*Gadis Indramayu*) was able to achieve—the Alliance's leader decided not to continue the same effort. It is likely that Dar's case will be the first and last for farmers seeking official government release of their own cultivars.

Following the Kediri cases, involving Kun and others, a group of scientists from the Bogor Agricultural Institute (*Institut Pertanian Bogor*, IPB) in collaboration with IPPTHI formed AB2TI to help farmers to save their own cultivars. Since the association is legally formed, it officially has the right to collect and preserve local seeds from various places in Indonesia at a time when the regency governments did not yet have any interest in preserving and documenting local varieties. One initiator, H. Roni (the leader of IPPHTI of the Indramayu Regency) said that: "[f]rom the time the film of *Bisa Dèwèk,* initiated by Winarto and anthropologists from UI was produced, Indramayu farmers considered how to institutionalize farmers' own cultivars and protect the farmer-plant-breeders." H. Roni claimed that this idea became the "seeds of the formation of AB2TI in 2012." Not only the criminalization and intimidation of farmer-plant-breeders, but also the success of the farmers with the judicial review motivated Professor Andreas Dwi Santosa from IPB to collaborate with farmers in creating a forum to help farmer-plant-breeders.

The association has the objective of facilitating and accommodating farmers' own initiatives and creativity in producing their own cultivars so as to establish farmers' sovereignty in relation to their seeds.[34] Santosa as the leader of the association produced the motto "We do not need help. We do not want to be disturbed" (*Kami tidak perlu dibantu. Kami tidak mau diganggu*). The declared aim represented by this motto is to protect the farmers from criminalization by the authorities as well as by transnational seed companies. Farmers from various regencies in Indonesia sent their local seeds to AB2TI. The association has a special office and

[34] On the related concept of 'food sovereignty' see McKeon, 2015, 77–81).

laboratory to preserve those seeds and produce its own cultivars. *Ex-situ* conservation and scientific plant-breeding is what AB2TI provides as opposed to the individual farmer-plant-breeding as practiced in Indramayu, Purbalingga, and Lampung. It is still in the early stages of training farmers to be plant-breeders through the PPB FFSs (Winarto, 2011; Winarto & Ardhianto, 2011a, 2011b). Even though AB2TI has also initiated collaboration with the government in certifying cultivars, none has been released so far. In the view of the government, the standard of selection has not been adequate for plant breeding.

Understanding that certifying seeds with the government would not be easy, AB2TI then decided to release the cultivars themselves outside the government's regulation. Their objective has been to develop their own market through AB2TI shops to cater for farmers' own products.

### 7.5.3 Coping with the Consequences of Climate Change

Rice is the most vulnerable crop in the midst of increasing temperatures in the world. That was a strong reminder given by the late agrometeorologist, Kees Stigter, when collaborating with Universitas Indonesia (UI) team in assisting farmers to cope better with the consequences of climate change (Winarto et al., 2018). Understanding the difficulties of breeding rice varieties resistant to increasing heat (with an increase in the midnight minimum temperatures), Stigter advised farmers—joining the SFSs in Indramayu and East Lombok regencies—to try to discover among the existing rice varieties the ones that proved to be resistant under particular climate conditions. Farmer-rainfall-observers did use the opportunities provided in the SFS to select, evaluate, and trial rice varieties (Winarto & Stigter, 2016; Winarto et al., 2018). Based on their daily rainfall measurements and agroecosystem observations, monthly evaluation of their data for the past 30 days (3x10 days) periods, yield evaluation, and more recently, farmer-field-experimentation with rice varieties, farmers have gradually learned which varieties would be best suited to a particular ecosystem and climate condition in a certain planting seasons.

It is interesting to note that there are similarities and differences in farmers' choices of resistant or more suitable varieties based on different ecosystems, planting seasons, and particular climate conditions. The Indramayu rainfall-stations were divided into 4 zones (North-West zone, South-West zone, North-East zone, and South-East zone) each with a specific ecosystem. In one monthly evaluation in December 2017, we found that the farmer-rainfall-observers in the South-East and South-West zones, which rely heavily on rainfall as the main water source, would prefer rice varieties according to the particular climate conditions of El-Niño, La-Niña, or Normal.[35] Like the farmers in these two zones, farmers in the most

[35] Indonesia's climate is determined by many regular oceanic and air flows with all kinds of variations that determine the start, length, end and distribution of rainfall in the rainy season. Climate change is due to changes occurring in the regular flows in the ocean and the air on which changing

North-Eastern area also rely on rainfall as the main water source. However, in the very strong El-Niño season when no water is available, some farmers avoid planting rice based on their experience of harvest failure in such an extreme climate condition in 2015. Though one high yielding variety, Ciherang, became a main variety across all climate conditions, during La-Niña they prefer to plant long-maturing and glutinous rice varieties to avoid the logging of the plants. The neutral climate condition provides greater choice for farmers. One plant-breeder prefers to use his own cultivar (*Bongi*). For farmers in the North-West zone with abundant water irrigation, different climate conditions do not affect the farmers' choice of rice varieties. The most important consideration is to plant the same variety—or a similarly maturing variety—in any one area of the rice fields to avoid severe infestation of pests/diseases. Those who have lower elevated fields with inundated water for quite a long period during the rainy season always choose rice varieties resistant to abundant water.

The farmers' learning of selecting rice varieties indicates the very diverse nature of rice ecosystems and climate conditions, and the benefits of such learning in improving the farmers' potential ability to cope with the increasing variability of the climate. Unfortunately, there are no indications that those in government circles have realized this potential in their efforts to improve food productivity and security.

### 7.5.4 Preventing Abuse by External Agents

Whatever potentials farmers have in their capacity as food producers, their position is vulnerable to abuse by those in more powerful positions. How can they protect themselves and prevent any abuse of their innovation and practices? Plagiarism of their cultivars by outside agents is one main thing to avoid. Regarding this problem, it helps to join an organization such as AB2TI and to develop a good standardized documentation of their seeds. In this way, any attempts to use their seeds could be monitored. Similarly, Arifin, who decided not to participate in the AB2TI movement, is very cautious with people who come and want to see his data and document his seeds. He asks them in detail for their reasons and objectives in seeing his documentation. He is clearly concerned to avoid any plagiarism.

The Indramayu farmer-plant-breeders decided not to register their seeds under the name of the Head of Regency. More recently, many heads of regencies have declared farmers' cultivars as the "regency's prominent local varieties" in accordance with Article 7 of Law No. 29 of 2000 on the Protection of Plant Varieties and with Government Regulation No. 13 of 2004 on the Naming, Registration and Use

---

variations also have their influence. However, once in a while the temperatures of the Pacific Ocean suddenly change very irregularly to a much warmer or colder (near) surface temperature situation, causing large changes in atmospheric airflows that cause huge climate effects. Indonesia becomes much drier in an El Niño situation and much wetter in a La Niña situation. After some time (shorter or longer) the situation will return to normal (Stigter, personal note, 2014).

of the Initial Variety for the Making of Essentially Derived Varieties, which puts provincial governors, regency heads and mayors in charge of the registration of "local varieties". 501 such local varieties were registered up to 2014 (Kanniah & Antons, 2017, 297). The farmers do not want their cultivars to become subject to the "right and ownership of the regency" and not to the "right and ownership of individual farmers". Therefore, the situation remains fragile for farmers when it comes to protecting their own cultivars.

## 7.6 Reform of the Seed Legislation: The 2019 Law on the Sustainable Agricultural Cultivation System

The new Law No. 22 of 2019 on the Sustainable Agricultural Cultivation System was signed into law by President Joko Widodo on 18 October 2019.[36] The new law shows that the Indonesian government has learned the lessons from the Constitutional Court case and the subsequent discussions.[37] While the preamble of the law of 1992 spoke of the important role of a progressive, efficient and strong agricultural sector in reaching the goal of national development, farmers were mentioned only indirectly as "human resources" (*sumberdaya manusia*), whose quality needed to be raised to shape the sector. The preamble of the new law also stresses development goals, but farmers now occupy a central position in such efforts. The sustainable agricultural cultivation system does not just have to be "advanced, efficient, strong and sustainable", but also contribute to food sovereignty (*kedaulatan pangan*) and take into account the carrying capacity of the ecosystem and climate change mitigation and adaptation. In the explanatory memorandum to the preamble, the goals of progress, efficiency and strength are now joined by a number of principles, which are also listed in Article 2. Many of these relate to concerns that were raised in the Constitutional Court case and in the discussions since. To give a few examples from the list: the principle of "usefulness" means that the system should bring the greatest possible use for the prosperity and quality of life of the people.[38] The principle of "sovereignty" means that the system has to be implemented with high respect for the sovereignty of farmers who have rights and freedoms in the framework of self-development.[39] The principle of "autonomy" means that the system has to be implemented in an independent manner prioritizing the capacity of domestic resources.[40]

---

[36] Available at http://peraturan.bpk.go.id/Home/Details/123688/uu-no-22-tahun-2019 (last accessed 29 January 2020).

[37] When introducing the law, the Ministry of Agriculture pointed out that it implemented the Constitutional Court decision, but regulated also subject matter that was not yet regulated in the previous law, such as fertilizer, pesticides and agricultural tools and machinery (Kementerian Pertanian Republik Indonesia, 2019).

[38] Explanatory note to Article 2.a.

[39] Explanatory note to Article 2 c.

[40] Explanatory note to Article 2 f.

The principle of "local wisdom" means that the implementation has to balance social, economic and cultural characteristics and the supreme values that are in force in a local social order.[41] There is further the principle of the "conservation of the function of the living environment" which means that the system has to use means, infrastructure, procedures and technologies which do not disturb the function of the living environment, whether in biological, mechanical, geological or chemical manners.[42]

However, while there is a clear shift towards sustainability goals, some of the developmental aspects that were unsuccessfully challenged in the Constitutional Court case on the previous law, have remained. As in the previous law, the freedom of farmers to choose plant and animal varieties and the method of their cultivation, is limited by the obligation to prioritise development plans,[43] although the planning authority of national and regional governments under the new decentralised system is not as wide as that of the centralised Suharto administration under the 1992 law.[44] An incentive to stay in line with such plans is provided in Article 11, which promises government assistance for those farmers (and presumably only for those farmers), who exercise their choices in accordance with national and regional development plans.

The new law prescribes a wider scope how to obtain seeds focusing on the invention and/or selection of superior varieties or traits and/or their introduction from outside (Article 25). Anyone can engage in breeding (Article 26(2)), but the search for and collection of genetic resources will be the responsibility of the central and regional governments and can be carried out by individual and legal persons that have been appropriately licensed. Small farmers are exempted from this licence requirement (Article 27(2)). Small farmers will also be exempted from the general requirement that all results of plant breeding have to be officially released by the government (Article 29(1). In both cases, however, small farmers will be required to report their activities to the government—to the central government via the regional government in the case of genetic resources (Article 27(3)) and to the regional government in the case of the results of plant breeding (Article 29(2)).[45] Importantly, varieties produced by small farmers can only be circulated in a limited way and within one regency (*kabupaten)* or city (*kota*) (Article 29(3)). Therefore, the new law implements the principle of making an exception for small farmers as required by the Constitutional Court decision, but, different from that decision, it narrowly circumscribes the geographical scope for this freedom. The wide freedom to distribute their varieties Indonesia wide that farmers had assumed after the decision, therefore, seems to have been curtailed.

---

[41] Explanatory note to Article 2 i.

[42] Explanatory note to Article 2j.

[43] See Article 10(2) Law No. 22/2019.

[44] Cf. Article 5 of the new law with Article 5 of the 1992 law, which even allowed the government to "regulate the production of certain plant cultivation in accordance with the national interest" (Article 5(1) c. Law No. 12/1992).

[45] Articles 27(3) and Article 29(2).

The concept of 'small farmer' has to be collected from the explanatory notes. An explanatory note of the draft law earlier circulated for discussion had included landless farmers that work on an area of at the most two hectares; landowning farmers with at the most two hectares of land; and horticultural farmers, gardeners and small-scale livestock breeders according to criteria in separate pieces of legislation.[46] By comparison, the explanatory note on Article 27(2) of the now enacted law defines "small farmers" as subsistence farmers.[47] Equating subsistence farmers with those working on two hectares or less is indeed a commonly used approach (Rapsomanikis, 2015, 1), but in principle a definition of subsistence farming is complicated, as the distribution of farm size depends on many factors and on agro-ecological and demographic conditions (Rapsomanikis, 2015, 2). Heidhues and Brüntrup (2003, 6),[48] discussing definitions in agricultural economics highlight three sources of ambiguity: subsistence can be used as a concept of market integration and/or as a measurement for standard of living; it can be measured from the viewpoint of consumption and/or production; and any subsistence indicator can be on a scale from almost 100% to practically zero. The definition using farm size in the draft law has now been discarded and there is no definition of subsistence farming in the new law. Since it is perfectly common for subsistence farmers to also trade their products and to participate in market transactions to some extent, there is some discretionary space as to who will be allowed to benefit from the small farmer exemptions. Such ambiguities create uncertainties for those relying on the provisions.

In spite of the declared intention to make matters easier for small farmers in particular, the new regulations nevertheless introduce new bureaucratic reporting requirements to the government in case small farmers search for and collect genetic resources (Article 27(3)) and release a "small farmers' breeding variety" (*varietas hasil pemuliaan petani kecil*) (Article 29(2). The details of these reporting requirement will be regulated further in a Government Regulation in case of the collection of genetic resources (Article 27(5), but there is no further regulation of the reporting process for the release of the farmers' variety. This again creates some uncertainty as to what farmers are expected to do. Ministry of Agriculture Regulation No. 40/Permentan/TP.010/11/2017, although not mentioned in the new law, requires that a 'small farmers' breeding variety' has to be registered with the implementing authority regarding food plants, plantations or animal husbandry respectively and has to be given a name that indicates the place of the breeding activity (Article 36(3) and (4) of Ministry of Agriculture Regulation No. 40/2017). As with the geographical scope

[46] See the explanatory note to Article 27(3) of the draft law, http://www.dpr.go.id/doksileg/proses2/RJ2-20171109-024008-6387.pdf (last accessed 4 February 2020). Ministry of Agriculture Regulation No. 40/Permentan/TP.010/2017 Regarding the Release of Plant Varieties also defines small farmers as operating on at the most 2 hectares and at the most 25 hectares in the case of a plantation (Article 36(2)).

[47] See the explanatory note to Article 27(2): "What is meant by "small farmers" are farmers, who work every day in the agricultural sector with yields that are only sufficient to meet the needs of daily life."

[48] See also Annex 1 in Heidhues and Brüntrup (2003: 18–21) for further details.

for the distribution discussed earlier, the reporting requirements bring new limitations for the freedom that farmers believe to have won after the favourable Constitutional Court decision.

A further ambiguity stems from the fact that seeds are also included in the "agricultural cultivation tools" regulated in Chapter XI of Law No. 22/2019. As such they have to comply with the safety and quality standards mentioned in Article 66(1). While these would normally be assessed in a certification procedure, small farmers and their products are exempted from this process (Article 66(2), (4). They are also exempted from technical minimum conditions prescribed by the Minister (Article 66(3), (4)) and from labelling requirements (Article 68(1), but in the current wording of the law not from the safety and quality standards as such. This is important, because violation of Article 66(5) attracts criminal penalties under Article 121.

It appears that farmers can also be operators of businesses (*pelaku* usaha) as defined by the law. Article 1 No. 21 defines a "business operator" as "any person who conducts a business related to agricultural infrastructure, tools, agricultural cultivation, crops, the time after the harvest, the processing and marketing of agricultural products as well as services supporting agriculture and domiciled in the territory of the Republic of Indonesia". Regulation of the Minister of Agriculture No. 39 of 2010 Concerning Guidelines for the Licensing of Food Crop Cultivation Businesses also includes small famers and small farmers with limited land in the definition of "food crop cultivation business operators". Given the emphasis on subsistence farming in Law No. 22/2019, it seems somewhat counterintuitive to regard small farmers as also operating businesses, but the current definition does not exclude them. This means that they could attract criminal penalties under Article 125, if they operate a business on a certain scale without a government permit. Clearly excluded, on the other hand, are business operations on collectively held land subject to the customary law of *adat* communities (Article 86(2)), unless there is an agreement with the community.[49] This provision can be seen as a reaction to yet another important Constitutional Court decision that opened the door for the recognition of the rights of *adat* communities to their customary lands.[50]

Apart from the criminal provisions discussed above that could become an issue for small farmers because of ambiguous wording used in other parts of the law, the exclusion of small farmers in the provisions on the release of varieties and the search for and collection of genetic resources now means that they no longer need to fear the immediate application of criminal provisions related to unauthorised acts in this regard. This relative level of certainty requires, however, that they stay with their distribution activities within their own district or city and that they are actually classified as small farmers. There remains some uncertainty with regards to the latter

[49] *Adat* is often loosely translated as 'customary law' (Utrecht & Djindang, 1983, 99), but its meaning is much wider (von Benda-Beckmann, 1979, 113–114). After constitutional reform brought the recognition of *adat* communities (*masyarakat adat*), it has become an important argument for forest-dwelling communities to claim recognition of their land rights (Lowenhaupt Tsing, 2009).

[50] See Constitutional Court case 35/PUU-X/2012 and the analysis in Rachman and Siscawati (2017).

classification, which depends on how the authorities will interpret the 'subsistence farming' requirement.

Finally, as is often the case with revised laws in Indonesia, the new law leaves all laws and regulations in force that were issued to implement Law No. 12 of 1992 and are not contradicting the new law (Article 129), although Law No. 12 of 1992 itself will be declared invalid (Article 130). As was shown above, because these regulations were issued prior to the Constitutional Court decision, they do not exempt small famers from their scope and some of them, therefore, contradict the new law. While this makes them legally invalid in relation to particular points, the Law No. 22/2019 unfortunately puts the burden to argue this on the small farmers that are benefitting from a different regulation. Therefore, small farmers may well continue to need the help of legally trained advisers in NGOs and farmers' associations in arguing with government and enforcement personnel about the extent of the legal changes.

Fortunately, and as discussed above, the important Ministry of Agriculture Regulation No. 40/Permentan/TP.010/11/2017 Regarding the Release of Plant Varieties clearly replaced the Ministry of Agriculture Regulation No. 61/Permentan/OT.140/10/2011 with its undifferentiated focus on "superior varieties" and varieties that are "distinct, uniform and stable". This Regulation No. 40/2017 also introduced the new term "small farmers' breeding variety", which it exempted in accordance with the Constitutional Court decision from the testing, evaluation, manner of release and manner of withdrawal prescribed for other varieties. In spite of this widespread exemption, this regulation already introduced a registration requirement for such farmers' varieties that anticipated the reporting requirements that have now been introduced by Law No.22/2019. Interestingly, the 2017 Ministry of Agriculture Regulation also distinguishes the 'small farmers' breeding variety' from the 'local variety', which is mentioned elsewhere in the Regulation as undergoing testing if proposed for release (Article 14). As discussed earlier, the 'local variety' is mentioned in Article 7 of the Plant Variety Protection Act of 2000 as being "owned by the community", but "controlled by the state". The implementing regulation has put local authorities in charge of representing the community in the registration process. When the Plant Variety Protection and Agricultural Permit Centre (*Pusat Perlindungan Varietas Tanaman dan Perizinan Pertanian*) of the Ministry of Agriculture published registrations for the period 2005 to May 2014 on its website, it showed that 468 local varieties had been registered by mayors, officers and governors of provinces (Antons, 2015, 470). The differentiation in the recent Ministry of Agriculture Regulation No. 40/2017 and in Law No. 22/2019 between these 'local varieties' and the new term of 'small farmers' breeding variety' allows the conclusion that 'local varieties' are understood as public domain varieties under the control of the state, while 'small farmers' breeding varieties' are the results of breeding activities of farmers that have difficulties to go through the "distinct, uniform and stable" testing, but should nevertheless be allowed and supported. In practice, of course, there may be significant overlap between the two, if farmers use 'local varieties' for these breeding activities.

## 7.7 Conclusion

Indonesia is one of the developing countries in Asia with an extraordinary lively debate about the future of the country's farming sector. The discussion is driven by a large number of NGOs representing the interests of farmers as well as by other NGOs in the field of human rights or international trade relations. The country has come a long way from the days of the authoritarian Suharto government with its top-down planning policies centralised in Jakarta. A few developments are particularly worth mentioning to explain this development. The initial changes came with what has been called in Indonesia the reformation process (*reformasi*) after the end of the Suharto years and four amendments to the Indonesian Constitution. Among other things, they brought a widespread decentralisation with distribution of power, taxes and income to the regions.[51] Benefitted from this have also the councils of villages and townships around the country that now have the funds to run workshops and projects on specific local matters in collaboration with one or several of the many NGOs.

Another important result of the reformation years with their constitutional reform is the inclusion of a chapter on human rights in the Indonesian Constitution[52] and the creation of the Indonesian Constitutional Court.[53] NGOs with lawyers and legal expertise, such as the Indonesian Human Rights Committee for Social Justice, as well as organisations representing the interests of particular sections of the population, such as the Indonesian Farmers' Union, are frequently contesting the constitutionality of government laws. Over the last decade, the Constitutional Court has made a number of progressive decisions recognising, for example, the rights of *adat* communities to their traditional land under certain circumstances[54] and it has acknowledged the long-denied existence in Indonesia of belief systems outside of the mainstream religions (Suryowati, 2017). The important Constitutional Court decision on farmers' rights regarding their breeding of plants and use of seeds has to be seen in this context of successful litigation aiming at the realisation of human rights promised in the revised Indonesian Constitution.

Governments have been following up on such landmark cases with legislation that aims at realising these rights. In the case of the victory of Indonesia's farmers with regards to their breeding and seed use activities, Ministry of Agriculture Regulation No. 40/Permentan/TP.010/11/2017 Regarding the Release of Plant Varieties was followed by Law No. 22 of 2019 on the Sustainable Agricultural Cultivation System. However, while both clearly pay attention to the exceptions that the Constitutional Court has required for the small scale farming sector, they again put some fetters on farmers' freedom to operate at the local level by requiring

---

[51] See, Chapter VI on Regional Government and Chapter VIIA on the Regional Parliament.

[52] See Chapter XA on Human Rights.

[53] See Article 24(C)of the revised Indonesian Constitution of 1945.

[54] See Constitutional Court decision No. 35/PUU-X/2012 of 26 March 2013 (Mahkamah Konstitusi, 2013b) and the analysis in Rachman and Siscawati (2017).

reporting of such activities to the relevant government agencies and, in the case of the Ministry of Agriculture Regulation, the registration with local authorities of 'small farmers' breeding varieties'. In terms of how widely farmers could circulate their material, the draft law will not allow them to exchange it beyond a limited territorial reach and this puts question marks behind the current practice to exchange plant material at fairs attended by farmers from many different districts and provinces or to market local varieties Indonesia-wide by using social media. For the exemptions to be effective, much will depend on how precisely 'small farmers' are defined and there is relatively little guidance in this matter. Other ambiguities in the law also continue to expose farmers' activities in an unnecessary manner to criminal provisions, even if the chance of application of these provisions in such cases is remote. Finally, it will be interesting to see how the newly introduced 'small farmers' breeding varieties' relate to the 'local varieties' introduced and protected in the country's Plant Variety Protection Act.

This research confirms the importance of local knowledge in selecting and breeding varieties suitable for conditions of climate change and environmental crisis.[55] In Indonesia as in other parts of the world, however, voices from farmers and scientists sympathetic to grassroots level concerns are struggling to be heard in discourses often dominated by the growth and output oriented ideology of developmental governments. These different visions for the future of agriculture in Indonesia are now increasingly debated in court and in both national and regional parliaments. The new laws and regulations show clearly that a better understanding of the innovative and creative contributions of small farmers has been established, even if they curtail to some extent the freedom obtained early in the decade in the Constitutional Court.

## References


Agrawal, A. (1995). Dismantling the divide between indigenous and scientific knowledge. *Development and Change, 26*, 413–439.

Ansori, S. (2011). Konservasi Benih oleh Petani Pemulia: Menuju Pembentukan Institusi? In Y. T. Winarto (Ed.), *Bisa Dèwèk: Kisah Perjuangan Petani Pemulia Tanaman* (pp. 227–306). Depok, Indonesia: Gramata Publishing.

Antaranews. (2019). *DPR Aceh sesalkan kriminalisasi kepala desa terkait bibit padi*. Antaranews.com, 24 July 2019, http://www.antaranews.com/berita/974504/dpr-aceh-sesalkan-kriminalisasi-kepala-desa-terkait-bibit-padi. Accessed 4 Feb 2020).

Antons, C. (2007). *Sui Generis* protection for plant varieties and traditional knowledge: The example of India. *European Intellectual Property Review, 29*(12), 480–485.

Antons, C. (2009). Indonesia. In P. Goldstein & J. Straus (Eds.), *Intellectual property in Asia: Law, economics, history and politics* (pp. 87–128). Berlin, Germany/Heidelberg, Germany: Springer.

Antons, C. (2010). *Sui Generis* protection for plant varieties and traditional knowledge in biodiversity and agriculture: The international framework and national approaches in the Philippines and India. *Indian Journal of Law and Technology, 6*, 89–139.


[55] For comparative observations regarding the usefulness of local knowledge in other countries see Coelho (2007) and the contributions in Ellen (2007).

Antons, C. (2015). Epistemic communities and the "people without history": The contribution of intellectual property law to the "safeguarding" of intangible cultural heritage. In I. Calboli & S. Ragavan (Eds.), *Diversity in intellectual property: Identities, interests, and intersections* (pp. 453–471). New York: Cambridge University Press.

Antons, C. (2016). Article 27(3)(b) TRIPS and plant variety protection in developing countries. In H. Ullrich, R. M. Hilty, M. Lamping, & J. Drexl (Eds.), *TRIPS plus 20: From trade rules to market principles* (pp. 389–414). Heidelberg, Germany/New York/Dordrecht, the Netherlands/ London: Springer.

Antons, C. (2017). Legal and cultural landscapes: Cultural and intellectual property concepts, and the 'safeguarding' of intangible cultural heritage in Southeast Asia. In C. Antons (Ed.), *The Routledge handbook of Asian law* (pp. 250–268). London/New York: Routledge.

Antons, C. (2019). Intellectual property in plant material and free trade agreements in Asia. In K.-C. Liu & J. Chaisse (Eds.), *The future of Asian trade deals and IP* (pp. 229–256). Oxford, UK: Hart Publishing.

Ardhianto, I. (2011). Benih Milik Saya, Kami, atau Mereka? Tumbuh Kembang oleh Petani Pemulia: Menuju Pembentukan Institusi? In Y. T. Winarto (Ed.), *Bisa Dèwèk: Kisah Perjuangan Petani Pemulia Tanaman di Indramayu* (pp. 307–325). Depok, Indonesia: Gramata Publishing.

Brush, S. (1993). Indigenous knowledge of biological resources and intellectual property rights: The role of anthropologists. *American Anthropologist, 95*(3), 653–686.

Brush, S. (1996). Whose knowledge, whose genes, whose rights? In S. B. Brush & D. Stabinsky (Eds.), *Valuing local knowledge: Indigenous people and intellectual property rights* (pp. 1–24). Washington, DC: Island Press.

Butt, S., & Lindsey, T. (2009). The people's prosperity? Indonesian constitutional interpretation, economic reform and globalization. In J. Gillespie & R. Peerenboom (Eds.), *Regulation in Asia: Pushing back on globalization* (pp. 270–295). London/New York: Routledge.

Cleveland, D. A., Soleri, D., & Smith, S. E. (2000). A biological framework for understanding farmers' plant breeding. *Economic Botany, 54*, 377–394.

Coelho, J. P. B. (2007). The state, the community, and natural calamities in rural Mozambique. In de Sousa Santos, Boaventura (Ed.), *Another knowledge is possible: Beyond northern epistemologies* (pp. 219–245). London/New York: Verso.

Cypher, J. M., & Dietz, J. L. (1997). *The process of economic development*. London/New York: Routledge.

de Sousa Santos, B., Nunes, J. A., & Meneses, M. P. (2007). Opening up the canon of knowledge and recognition of difference. In de Sousa Santos, B. (Ed.), *Another knowledge is possible: Beyond northern epistemologies*, xix–lxii. London/New York: Verso.

Dove, M. R. (2000). The life-cycle of indigenous knowledge, and the case of natural rubber production. In R. Ellen, P. Parkes, & A. Bicker (Eds.), *Indigenous environmental knowledge and its transformations: Critical anthropological perspectives* (pp. 213–251). London/New York: Routledge.

Ellen, R. (Ed.). (2007). *Modern crisis and traditional strategies: Local ecological knowledge in island Southeast Asia*. New York/Oxford, UK: Berghahn Books.

Fox, J. J. (1991). Managing the ecology of rice production in Indonesia. In J. Hardjono (Ed.), *Indonesia: Resources, ecology, and environment* (pp. 61–84). Singapore, Singapore: Oxford University Press.

Fox, J. J. (2014, March). Fast breeding insect is devastating Java's rice – Thanks to pesticides. *Jakarta Globe, 7*, 2014.

Frossard, D. (2005). In field or freezer? Some thoughts on genetic diversity maintenance in rice. In M. R. Dove, P. E. Sajise, & A. A. Doolittle (Eds.), *Conserving nature in culture: Case studies from Southeast Asia* (pp. 144–166). New Haven, CT: Yale Southeast Asian Studies.

GRAIN & Sharma, D. (2005). India's new seed bill. *Seedling,* July 2005, 22–27.

Harian Aceh. (2019). *Moeldoko Respon HKTI Yang Siap Bantu Kades Ditahan di Aceh.* 27 July 2019. http://www.harianaceh.co.id/2019/07/27/moeldoko-respon-hkti-yang-siap-bantu-kades-ditahan-di-aceh. Accessed 4 Feb 2020.

Heidhues, F., & Brüntrup, M. (2003). Subsistence agriculture in development: Its role in processes of structural change. In S. Abele & K. Frohberg (Eds.), *Subsistence agriculture in Central and Eastern Europe: How to break the vicious circle?* (pp. 1–27). Halle, Germany: Institute of Agricultural Development in Central and Eastern Europe.

Indonesian Human Rights Committee for Social Justice. (2013). *Our seed, our sovereignty: Seed law victory in Indonesia.* https://www.grain.org/en/article/4774-our-seed-our-sovereignty-seed-law-victory-in-indonesia. Accessed 5 Feb 2020.

Janis, M. D., Jervis, H. H., & Peet, R. (2014). *Intellectual property laws of plants*. Oxford, UK: Oxford University Press.

Jhamtani, H., & Patria, D. (2006). Case documentation: Indonesian farmers prosecuted for breeding their own seeds. *GRAIN*., 24 October. https://www.grain.org/en/article/2208-indonesian-farmers-prosecuted-for-breeding-their-own-seeds. Accessed 5 Feb 2020.

Kanniah, R., & Antons, C. (2012). Plant variety protection and traditional agricultural knowledge in Southeast Asia. *Australian Journal of Asian Law, 13*(1), 1–23.

Kanniah, R., & Antons, C. (2017). The regulation of innovation in agriculture and sustainable development in India and Southeast Asia. In C. Antons (Ed.), *The Routledge handbook of Asian law* (pp. 287–309). London/New York: Routledge.

Kementerian Pertanian Republik Indonesia (2019). *RUU Sistem Budidaya Pertanian Berkelanjutan, Berpihak kepada Petani Kecil*. 25 September 2019. https://www.pertanian.go.id/home/?show=news&act=view&id=3987. Accessed 4 Feb 2020.

Kompas. (2019). *Ini Alasan Polisi Tahan Inovator Benih IF8*. Kompas.com, 29 July 2019. https://regional.kompas.com/read/2019/07/29/09444071/ini-alasan-polisi-tahan-inovator-benih-if8. Accessed 4 Feb 2020.

Lewis, W. A. (1954). Economic development with unlimited supplies of labour. *Manchester School, 22*(May), 139–191.

Lowenhaupt Tsing, A. (2009). *Adat*/indigenous: Indigeneity in motion. In C. Gluck & A. L. Tsing (Eds.), *Words in motion: Towards a global lexicon*. Durham, UK/London: Duke University Press.

Lyon, F. (1996). How farmers research and learn: The case of arable farmers of East Anglia, UK. *Agriculture and Human Values, 13*(4), 39–47.

McKeon, N. (2015). *Food security governance: Empowering communities, regulating corporations*. London/New York: Routledge.

Nafziger, E. W. (2006). *Economic development*. New York: Cambridge University Press.

Putra, A. (2019). Mengapa Tgk Munirwan Ditahan, Kadis: Semata-mata untuk Melindungi Masyarakat. *Beritakini*, 25 July 2019. https://beritakini.co/news/mengapa-distanbun-aceh-polisikan-tgk-munirwan-kadis-semata-mata-untuk-melindungi-masyarakat/index.html. Accessed 5 Feb 2020.

Rachman, N. F., & Siscawati, M. (2017). Forestry law, *masyarakat adat* and struggles for inclusive citizenship in Indonesia. In C. Antons (Ed.), *The Routledge handbook of Asian law* (pp. 224–249). Oxford, UK/New York: Routledge.

Rapsomanikis, G. (2015). *The economic lives of smallholder farmers: An analysis based on household data from nine countries*. Rome: Food and Agriculture Organization of the United Nations.

Republika. (2019). *HKTI Siap Bantu Kades yang Ditahan Akibat Kasus Benih*. Republika.co.id, 28 July 2019. https://nasional.republika.co.id/berita/pvbjmq349/hkti-siap-bantu-kades-yang-ditahan-akibat-kasus-benih. Accessed 4 Feb 2020.

Santosa, D. A. (2013). Kemenangan Kaum Tani. *Kompas*, August 6, 2013.

Silitoe, P. (2006). Ethnobiology and applied anthropology: Rapprochement of the academic with the practical. In R. Ellen (Ed.), *Ethnobiology and the science of humankind* (pp. 147–175). Malden, MA/Oxford/Carlton, VIC/London: Blackwell Publishing.

Simmonds, N. W. (1979). *Principles of crop improvement*. London: Longman Group Ltd..

Stigter, C. (K.). J. (2016a). Climate prediction for farmers: Should we laugh or should we cry? In Y. T. Winarto (Ed.), *Krisis Pangan dan 'Sesat Pikir': Mengapa Masih Berlanjut?* (pp. 21–32). Jakarta, Indonesia: Yayasan Pustaka Obor Indonesia.

Stigter, C. (K.). J. (2016b). Climate crises in the livelihood of Indonesian rice farmers. In Y. T. Winarto (Ed.), *Krisis Pangan dan 'Sesat Pikir': Mengapa Masih Berlanjut?* (pp. 33–40). Jakarta, Indonesia: Yayasan Pustaka Obor Indonesia.

Suryowati, E. (2017). Putusan MK Membuat Eksistensi Penghayat Kepercayaan Diakui Negara. *Kompas*, November 7, 2017. https://nasional.kompas.com/read/2017/11/07/18573861/putusan-mk-membuat-eksistensi-penghayat-kepercayaan-diakui-negara. Accessed 12 Nov 2018.

Tempo. (2005). *Penjara Bagi Petani Kreatif.* August 28, 2005. https://nasional.tempo.co/read/65842/penjara-bagi-petani-kreatif. Accessed 27 Feb 2018.

UNCTAD-ICTSD. (2005). *Resource book on TRIPS and development*. Cambridge, UK/New York: Cambridge University Press.

Utomo, F. (2013). *Bersemi Dalam Tekanan Global: Kriminalisasi Petani, Inisiatif Benih Lokal, & Uji Materi UU No. 12/1992 tentang Sistem Budidaya Tanaman*. Jakarta, Indonesia: Yayasan FIELD Indonesia.

Utrecht, E., & Djindang, M. S. (1983). *Pengantar dalam Hukum Indonesia*. Jakarta, Indonesia: Penerbit Sinar Harapan.

von Benda-Beckmann, F. (1979). *Property in social continuity: Continuity and change in the maintenance of property relationships through time in Minangkabau, West Sumatra*. The Hague, the Netherlands: Martinus Nijhoff.

Winarto, Y. T. (2004). *Seeds of knowledge: The beginning of integrated pest management in Java*. New Haven, CT: Yale Southeast Asia Council.

Winarto, Y. T. (Ed.). (2011). *Bisa Dèwèk: Kisah Perjuangan Petani Pemulia Tanaman di Indramayu*. Depok, Indonesia: Gramata Publishing.

Winarto, Y. T. (2011a). Kembalinya Benih dan Pengetahuan Lokal dalam Budi Daya Padi. In Y. T. Winarto (Ed.), *Bisa Dèwèk.: Kisah Perjuangan Petani Pemulia Tanaman di Indramayu* (pp. 201–229). Depok, Indonesia: Gramata Publishing.

Winarto, Y. T. (2011b). Bisa Dèwèk: Menuju Etnografi Komplisitas Suatu Dinamika. In Y. T. Winarto (Ed.), *Bisa Dèwèk: Kisah Perjuangan Petani Pemulia Tanaman di Indramayu* (pp. 1–22). Depok, Indonesia: Gramata Publishing.

Winarto, Y. T., & Ardhianto, I. (2011a). Tumbuh Kembang Benih-benih Pemuliaan Tanaman Padi di Ladang Petani. In Y. T. Winarto (Ed.), *Bisa Dèwèk: Kisah Perjuangan Petani Pemulia Tanaman di Indramayu* (pp. 153–200). Depok, Indonesia: Gramata Publishing.

Winarto, Y. T., & Ardhianto, I. (2011b). Mencatat Jejak Pengetahuan dalam Praktik dan Produk Pemuliaan Tanaman. In Y. T. Winarto (Ed.), *Bisa Dèwèk: Kisah Perjuangan Petani Pemulia Tanaman di Indramayu* (pp. 230–276). Depok, Indonesia: Gramata Publishing.

Winarto, Y. T., & Stigter, K. (2016). Incremental learning and gradual changes: 'Science field shops' as an educational approach to coping better with climate change in agriculture. In L. Wilson & C. Stevenson (Eds.), *Promoting climate change awareness through environmental education* (pp. 60–95). Hershey, PA: IGI Global.

Winarto, Y. T., Stigter, C. J., & Wicaksono, M. T. (2017). Transdisciplinary responses to climate change: Institutionalizing agrometeorological learning through science field shops in Indonesia. *Austrian Journal of South-East Asian Studies, 10*(1), 65–82.

Winarto, Y. T., Walker, S., Ariefiansyah, R., Prihandiani, A. F., Taqiuddin, M., & Nugroho, Z.C. (2018, January). *Institutionalizing science field shops: Developing response farming to climate change*. Case Study. http://www.fao.org/3/I8454EN/i8454en.pdf. Accessed 10 May 2018.

## Website

Balai Litbang Pertanian. Regulasi: Peraturan Mentan No. 61/Permentan/OT.140/10/2011 Tahun 2011. http://www.litbang.pertanian.go.id/regulasi/one/19/. Accessed 21 Dec 2018.

Balai Besar Penelitian Tanaman Padi. (2016). *Tahukah Anda Kelas Benih Padi?* http://bbpadi.litbang.pertanian.go.id./index.php/info-berita/tahukah-anda/tahukah-anda-kelas-benih-padi. Accessed 10 May 2018.

Balai Besar Penelitian Tanaman Padi. http://bbpadi.litbang.pertanian.go.id/index.php. Accessed 21 Dec 2018.

Dewan Perwakilan Rakyat. 2018. *Rancangan Undang-Undang Tentang Sistem Budidaya Pertanian Berkelanjutan*. Draft Law. http://parlemen.net/wp-content/uploads/2018/03/Naskah-RUU-Sistem-Budidaya-Pertanian-Berkelanjutan-12Feb18.pdf. Accessed 8 Nov 2018.

## Film Production

Bisa Dèwèk (We Can Do It Ourselves). (2007). Ethnographic film directed by Rhino Ariefiansyah, Haji Masroni and Warsiyah, produced by Indonesian Farmers Alliance for Integrated Pest Management of Indramayu Regency (IPPHTI) and Undergraduate Study Program, Department of Anthropology, Faculty of Social and Political Sciences, Universitas Indonesia.

## Court Decisions

Mahkamah Konstitusi. (2013a). Decision No. *99/PUU-X/2012 of*, (9 July 2013).

Mahkamah Konstitusi. (2013b). Decision No. *35/PUU-X/2012 of*, (26 March 2013).

# Chapter 8
# Development of Local Rice on the Tabanan Regency of Bali

**Ida Ayu Astarini, Made Pharmawati, Made Ria Defiani, and Kadambot H. M. Siddique**

**Abstract** This chapter surveys the cultivation of local rice varieties in the nine regencies of Bali, with a primary focus on the Tabanan Regency. It discusses the farming system of the region and the way in which rice is marketed. The chapter looks at the cultural role of traditional rice cultivation in Bali. It discusses the way in which farmers' rights under the International Treaty on Plant Genetic Resources for Food and Agriculture has been adopted by Indonesian legislation. The chapter also looks at the contribution which rice marketing makes to the Balinese agricultural sector, particularly export opportunities. The contribution which the replacement of white rice by red rice to health and nutrition is examined and the research on local rice in Bali is reviewed, including research into genetic erosion, drought stress and the response of rice genotypes to zinc fertilizer and mutation breeding studies on several Bali local rice varieties. The chapter concludes with a consideration of the potential for agriculture-based tourism in Bali.

**Keywords** Traditional rice cultivation in Bali · Red rice · Genetic erosion · Drought stress · Research into Balinese rice – agritourism

---

I. A. Astarini · M. Pharmawati · M. R. Defiani
Biology Department, Faculty of Mathematics and Natural Sciences, Udayana University, Bali, Indonesia

K. H. M. Siddique (✉)
Hackett Professor of Agriculture Chair and Director of The UWA Institute of Agriculture, University of Western Australia, Crawley, WA, Australia
e-mail: kadambot.siddique@uwa.edu.au



M. Blakeney, K. H. M. Siddique (eds.), *Local Knowledge, Intellectual Property and Agricultural Innovation*, https://doi.org/10.1007/978-981-15-4611-2_8

## 8.1 Introduction

Rice (*Oryza sativa* L.) is an annual plant that belongs to the family Poaceae. In Indonesia, rice is the primary source of carbohydrates. The country's population reached 262 million in 2017, with an average rice consumption of 111.58 kg/capita/year. In 2018, total rice consumption was estimated at around 29.57 million tons (BPS).

There are three main Asian rice varieties, based on Glaszmann's (1987) classification: (1) Indica, commonly grown in tropical countries, (2) Javanica, commonly grown in Java, Bali and Lombok, and (3) Yaponica/Sub-Yaponica, commonly grown in sub-tropical areas. Indica and Javanica varieties are commonly grown in Indonesia (Muhamad, Ebana, Fukuoka, & Okuno, 2017; Silitonga, 2004).

Traditional rice production in Indonesia has been passed down many generations, resulting in diverse Indonesia landraces (Muhamad et al., 2017). Rice was introduced to Indonesia from South China between 2000 and 1500 BC (Diamond & Bellwood, 2003) and from India about 2000–3000 years ago (Tanaka, 1998). There are about 17,000 rice accessions in Indonesia; 10,000 have been collected and documented, with 3500 characterized and used as parent lines to improve rice varieties within and outside Indonesia (Las, Suprihatno, Daradjat, Suwarno, & Satoto, 2004; Suhartini, 2010). Such high diversity is an unlimited resource for breeding and improvement of new and superior varieties.

Bali is one of 34 provinces of Indonesia—with a relatively small agricultural area compared to other provinces—that is spread across nine regencies (Table 8.1). Rice productivity in Bali is generally higher than other provinces in Indonesia due to good farm management and the use of an irrigation system and organization called 'Subak' (Suharyanto, Mulyo, & Widodo, 2015).

The presence of local rice varieties gradually declined in Bali after the Indonesian government introduced selected varieties in the 1970s, during the 'Green Revolution' era. At that time, farmers had to plant varieties designated by the government,

**Table 8.1** The nine regencies of Bali and their land use for agriculture and rice paddies

| | Land use (Hectares) | | |
|---|---|---|---|
| Regency | Total agricultural land | Irrigated rice paddy | Non irrigated rice paddy |
| Jembrana | 32,481 | 6289 | 469 |
| Tabanan | 62,216 | 21,089 | – |
| Badung | 28,067 | 9847 | 91 |
| Gianyar | 26,883 | 14,320 | – |
| Klungkung | 23,125 | 3779 | – |
| Bangli | 458,797 | 2876 | – |
| Karangasem | 60,165 | 7107 | 15 |
| Buleleng | 125,700 | 10,270 | 65 |
| Denpasar | 2919 | 2409 | – |
| Bali (total) | 407,534 | 77,986 | 640 |

Source: BPS – Statistics of Bali Province, 2017. Last update 17 January 2019

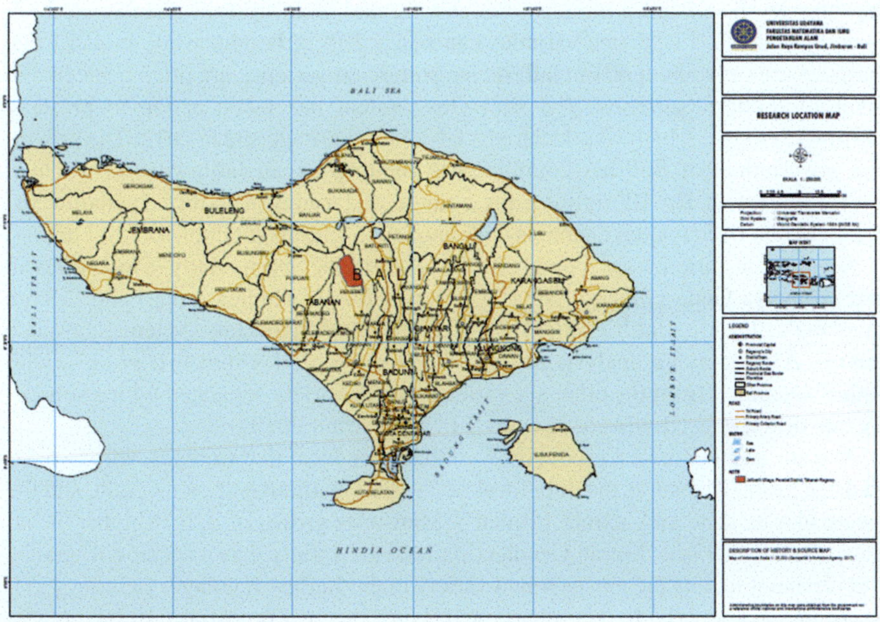


**Fig. 8.1** Map of Bali. The red area shows the location of Jatiluwih Village

including 'Peta', 'Ciherang', 'Cibogo', 'Inpari' and 'IR 64'. The farming system also had to be in accordance with governmental procedures, with high inputs such as chemical fertilizer and pesticides. In some areas, such as the Penebel District in the Tabanan Regency, the introduced varieties did not grow successfully; and therefore government allowed farmers in Penebel to grow only local varieties.

Jatiluwih Village is a rice production area in the Tabanan Regency (Fig. 8.1). Farmers in this village grow several local rice varieties that have been farmed for generations as they are well-adapted to the hilly areas (~700 m above sea level). Until now, the local Jatiluwih community prefers high quality rice, particularly red rice variety, which has doubled in price to standard varieties, about Rp 20,000–30,000 per kilogram.

Sitaresmi, Wening, Rakhmi, Yunani, and Susanto (2013) identified several local rice germplasms in Indonesia with resistance to biotic stresses—including ganjur or wood-mason (*Orseolia oryzae*) pest, bacterial leaf blight (*Xanthomonas oryzae* pv. oryzae), bacterial red leaf blight, leaf blast (*Pyricularia grisea*), neck blast, white streak leaves, wereng coklat (*Nilaparvate lugens*), and tungro (*Bacilliform virus*)—and abiotic stresses, such as drought, high Al, high Fe, salinity, low temperature, and shade.

Local rice varieties are at a disadvantage due to their longer growing periods and lower productivity than the modern varieties that are commonly grown in other areas of Bali, and Indonesia in general. Local rice varieties are usually harvested 6 months after sowing, with average productivity of 4 t/ha, while commercial varieties are harvested 3 months after sowing and yield up to 7 t/ha (Suwarno & Soenarjo, 2001). Local rice varieties can reach 2 m in height, while modern rice varieties are only about 60 cm tall. As a result, local varieties are often neglected in favor of modern varieties. The Bali Government has taken action to preserve Jatiluwih Village, which was declared a UNESCO heritage site in 2012. The provincial government of Bali has committed to funding Subak Jatiluwih in particular, with a budget of Rp 100 million for a range of farming activities, including the management of irrigation channels and renovation of Subak temples.

Seven local varieties are found in the Penebel district, particularly in the Jatiluwih and Wongaya Gede villages: 'Mansur', 'Ketan Beton' (Indica group) and 'Merah Cendana', 'Injin', 'Putih Cempaka', 'Ketan Tahun', and 'Jaka Selem' (Javanica group). A phenotypic analysis divided the seven varieties into two groups, with 'Injin' separated from the other six varieties. Each variety separated with a similarity rate of 62.16% (Budiwati, Kriswiyanti, & Astarini, 2019).

Mansur has the highest productivity of the seven local rice varieties, with 5–6.5 t/ha dried grain. Mansur is the only local rice variety without hair on its grain, and the locals like its taste and texture ('pulen'). Mansur is grown in Bali at higher elevations (BPTP, 2017a). 'Merah Cendana', a red rice variety, has a specific fragrance and is palatable; it is the most-farmed variety in the Jatiluwih village, yielding 4–5 t/ha. In 2017, about 10 ha were harvested, but there is a potential farming area of 223 ha (BPTP, 2017b).

Table 8.2 shows some morphological differences between local rice varieties in the Tabanan Regency, Bali (Budiwati et al., 2019).

## 8.2 Farming System

Subak Jatiluwih is a rice paddy region located on the hilly slopes of the Jatiluwih Village in the Penebel District of the Tabanan Regency. This area is 500–1000 m above sea level, with 303 ha of cultivated rice paddy, divided into seven 'tempek' or sub-irrigation groups/systems. Water resources come from spring water in the surrounding mountainous areas of Mount Batukaru, Mount Sanghyang and Mount

**Table 8.2** Agronomic characteristics of local rice in the Tabanan Regency of Bali (Budiwati et al., 2019)

| Variety | Number of shoots | Plant height (cm) | Time to harvest (days) | Yield (t/ha) |
|---|---|---|---|---|
| Mansur | 21 | 148 | 168 | 3.5 |
| Merah Cendana | 14 | 150 | 158 | 3.0 |
| Injin | 11 | 131 | 158 | 3.5 |
| Putih Cempaka | 16 | 152 | 158 | 3.0 |
| Ketan Beton | 18 | 144 | 163 | 2.0 |
| Ketan Tahun | 12 | 154 | 165 | 2.0 |
| Jaka Selem | 11 | 158 | 168 | 2.0 |

Adeng. Farmers plant rice twice per year; the first planting season starts at the end of December/early January, where farmers plant a local rice variety that is harvested by June, and the second season usually starts in July, where farmers plant a white rice variety that is harvested in 3.5–4 months (Ngadi, 2013).

Rainfall is important for agriculture in general and rice farming in Bali, to complement the irrigation system as in the rice paddy region. The average rainfall needed for optimum rice growth is 200 mm/month or 1500–2000 mm/year, distributed across 4–6 months. Good amounts of rainfall will benefit irrigation, as rice paddies need to be waterlogged during vegetative growth (Hasanah, 2007). Average temperatures in the Tabanan Regency range from 23 to 26 °C in the morning and 30 to 33 °C during the day. In general, rice needs 11–25 °C to germinate, 22–23 °C to flower, and 20–23 °C to develop seeds (Hasanah, 2007).

In Jatiluwih, farmers produce their own seed of the local rice variety (Fig. 8.2); the most popular local variety planted in the first season, started early January each

**Fig. 8.2** One-month-old seedlings ready to be transplanted into rice fields

year was the red rice Merah Cendana. Seeds are selected from the previous season's rice paddy population that had the best grain quality and preferred taste (Sitaresmi et al., 2013; Made Rehan, pers. comm.).

In the second planting season, farmers plant shorter growing season varieties, such as Mansur. The seeds are bought from the local market or nearby co-operation that provide seed and agricultural supplies (Ngadi, 2013; Subagyo, 2012).

Farmers in the Jatiluwih Village maintain their traditional farming system. Farmers prepare their land manually using 'cangkul', 'bajak', or a simple plowing machine (Fig. 8.3). Plant spacings for local rice in this region is 20 × 25 cm to maximize the benefits from sunlight, water, and nutrients for photosynthesis and maximum growth.

Common weeds found in local variety rice paddies in Tabanan regency are grouped into sedges and broad leaves. Sedges include Kambo mancik (*Scirpus juncoides* Roxb.), Adas – adasan (*Fimbristylis miliacea* (L.) Vahl.), Rumput Teki (*Cyperus rotundus* L.), Jeungan (*Cyperus difformis* L.), and *Oleocharis congesta* D. Don.). Broad leaves include Genjer (*Limnocharis* flava L.), *Commelina diffusa* Burm. f., *Lakum air* (*Ludwigia octovalvis* (Jacq.) Raven), and Eceng padi (*Monochoria vaginalis* (Burm. f.) C. Presl). Local bird pests include Pipit/Bondol jawa (*Lonchura leucogastroides*, Bondol haji (*Lonchura maja*), Bondol peking (Petingan) (*Lonchura punctulata*), Bondol hitam (*Lonchura ferruginosa*), and Manyar padi (*Ploceus manyar*). Local insect pests include Sundep dan Beluk, Wereng coklat (*Nilaparvata luges* Stal), and Wereng hijau, as well as caterpillars

**Fig. 8.3** Simple plowing machine used for land preparation

**Fig. 8.4** (**a**) 'Anggapan', a traditional cutter for harvesting local rice; (**b**) Farmer showing how to use 'anggapan'

**Fig. 8.5** Harvested red rice ready to be stored in the household

such as Ulat grayak (*Spodoptera litura* F.), Ulat separat (*Mythimna separate*), and Ulat tanduk hijau (*Melanitis leda ismene*) (Budiwati et al., 2019).

The local red rice variety is harvested with a tool called 'anggapan' (Indonesia: 'ani - ani', traditional cutter) (Fig. 8.4). Only the rice stalks are harvested with the 'anggapan', resulting in a longer harvesting period. The rice stalks are packed traditionally (Fig. 8.5) for storage in the household. The harvest is completed with the help of other local farmers. Modern cultivars are harvested differently, using a machine that harvests the rice stalks and leaves at the same time, followed by grain separation using a special machine. Workers come from outside Bali, mostly from East Java, to assist with the harvest.

Harvested local rice varieties in the Jatiluwih village are not generally sold but kept to fulfil the family's need for rice. Harvested rice can be stored for up to 1 year.

Extended periods of storage enhance the color the inside grain, particularly in red rice. Local people have their own rice storage house called 'Lumbung', which is used to store rice both for personal use and to be sold if harvested in a season is more than enough for family need.

## 8.3 Cultural Value

Rice farmers in the Jatiluwih Village practice traditional Balinese culture when farming rice. The concept of 'Tri Hita Karana' is passed from one generation to the next. In Balinese culture, Tri Hita Karana involves maintaining a balance between humans and the environment (in this case, rice paddy and the ecosystem), humans with humans, and humans with God. Farmers conduct ceremonies for every step in rice farming, starting with 'mapag toya' (the distribution of water into farmer's rice paddy), 'ngendag' (irrigating the land, preparing the soil), 'ngurit' (sowing the seed), 'nandur' (planting time), and 'mebiyukungkung' (harvest time).

Several ceremonies related to Balinese culture demand the use of local rice varieties. The 'Penjor', a Balinese pole used in ceremonies (Fig. 8.6a), is made with the stalks of local rice varieties as they are stronger and longer than those of introduced rice. Local rice varieties are in high demand, particularly during the 'Galungan' celebrations that occur every 6 months, according to the Balinese calendar. Galungan is the biggest cultural celebration in Balinese community, marking the winning of goodness (Dharma) over badness (Adharma). Local rice stalks are also used in temple ceremonies (Fig. 8.6b, c) and cremation ceremonies, particularly for 'ceg–ceg' to show the way to the cremation place, or the gate to heaven. Different colored rice (white, red, black) is routinely used in 'Atma Wedana', a ceremony to purify the souls of people that have passed away.

**Fig. 8.6** (**a**) Penjor along the road; (**b**) Barong made of local rice; (**c**) Local rice as part of a traditional ceremony

## 8.4 Farmers' Rights and the Protection of Local Rice

Farmers' rights to local germplasms for food is recognized in the International Treaty on Plant Genetic Resources for Food and Agriculture. The treaty was adopted by the Thirty-First Session of the Conference of the Food and Agriculture Organization of the United Nations on 3 November 2001. The treaty recognizes the enormous contribution farmers have made to the ongoing development of the world's wealth of plant genetic resources. It also states that farmers have the right to store, use, exchange, and sell agricultural seed, and that local governments are responsible for regulating priority needs in local agriculture (FAO, 2020). The international treaty has been adopted into Indonesian treaty, 'Undang – Undang Republik Indonesia no 4 Tahun 2006', in an effort to protect Indonesian farmer's right (UU no 4, 2006).

In Bali, local rice farmers use their own seed under the supervision of the Department of Monitoring and Seed Certification for Cereal and Horticultural Plants (Balai Pengawasan dan Sertifikasi Benih Tanaman Pangan dan Hortikultura, BPSBTPH).

In 2016, the Department of Agriculture's division on Agricultural Technology Research (BPTP) announced a program called 'self-sufficient seed village' in Bali to allow farmers share in within their village. The Department also supervised Subak Guama (an organization of farmers in Guama Village, Tabanan Regency) in their application to use new technology and blend traditional management with modern methods, beside managing the irrigation system for rice paddy in that village.

Subak Guama has established a co-operation named KUAT (Koperasi Usaha Agribisnis Terpadu, meaning Co-operation on Integrated Agribusiness) in 2001, which currently has 544 farmer members. Business activities undertaken through this co-operation include selling calves, agricultural production needs, and providing rice seeds.

Farmer organizations under Subak are very strong in Bali for managing the water distribution, planting times, and varieties needed by Subak members. Everything is agreed upon at member meetings, with the head of the Subak ('Pekaseh') playing an important role. Governments or companies wishing to collaborate with farmers for seed provision or other activities must approach the Pekaseh (Sutami, Londra, & Suastika, 2016).

Indonesia, as a member of the World Trade Organization, adopted the UPOV Convention treaty, as Undang-Undang No. 29 Year 2000 on Plant Variety Protection (UU Perlindungan Varietas Tanaman, for short). Most farmers are not aware of this treaty so local governments need to educate farmers about their rights and variety protection.

Plant local varieties are genetic resources that must be conserved. This can be done by registering the variety with the Center for Plant Variety and Agricultural Licensing Protection of Republic Indonesia. BPTP Bali assists in the registration process. Recently registered local varieties as germplasm include padi 'Gondrong

Sudaji', padi beras merah 'Munduk', 'padi ketan 'Gundil Sudaji', and padi 'Cicih Gundil' from Buleleng Regency, Bali (BPTP Bali, 4 July 2019).

## 8.5 Marketing of Local Rice

The agricultural sector is the second highest contributor to the income of Bali, after tourism. In 2018, agriculture contributed 13.81% compared with 23.34% for hotel and restaurant (BPS Bali Province, 2019).

Local rice varieties in Bali are mainly produced to meet local demand. In 2018, the Governor of Bali issued a new regulation, Pergub no 99 year 2018, for the marketing and use of local produce from agriculture, fisheries and industry. With the issue of this Pergub, the Head of the Department of Agriculture, Horticulture and Plantation stated that hotels and restaurants in Bali were obliged to buy local products for their businesses, with reasonable prices paid to farmers (Jarakpos 24/10/2019) (Fig. 8.7).

The demand for organic rice has increased recently, from 36 ton in 2013 to 60 tons in 2017. This has resulted in Jatiluwih farmers shifting to organic farming to meet the demand, particularly for local red rice varieties. CV Jatiluwih is one of the small companies that produces an organic red rice variety branded as 'Fragrant Red Rice'. Local farmers supply this rice has established a group called Kelompok Tani Beras Merah Organik Jatiluwih or Jatiluwih organic red rice farmer's group, which has been certified organic by LeSos. LeSos is an organization that has the mandate

**Fig. 8.7** Local rice products ready to sell in the Jatiluwih area

from Organic Food Competency Authority; Ministry of Agriculture, Indonesia to certify organic food. In 2017, CV Jatiluwih produced 50 tons of organic red rice, and is yet to meet the 60 ton demand. Fragrant Red Rice is marketed solely to the Tiara Dewata Supermarket in Denpasar, a well-known grocery store in Bali.

## 8.6 Export Opportunities

Red rice varieties offer excellent opportunities for export due to their health benefits; red rice has a lower carbohydrate content than other rice so it is more suitable for people with diabetes. It can be marketed as rice or rice flour. The demand for export to several countries, including the USA, Japan, and Canada, cannot be met as farmers are unable to produce sufficient amounts. Sowing to marketing takes 12 months, including a 5 month growing period, 4 months of storage in a special storage house ('Lumbung'), followed by packing and marketing. The value of red rice in shops is around Rp 20.000 per kg, or double that of white rice. Red rice exports to the USA reached 25 tons in 2018 (NusaBali.com, 2018).

## 8.7 Alternative Processing of Red Rice

Rice is the staple food for people in southern and eastern Asia, mostly in the form of white rice. White rice is mainly composed of starch and small amounts of protein, fat, and fiber. Studies have shown that a high intake of white rice is associated with an increase in the risk of type 2 diabetes and other metabolic disorders (Bahadoran, Mirmiran, Delshad, & Azizi, 2014; Hu, Pan, Malik, & Sun, 2012). In 2018, the prevalence of diabetes in Indonesia based on doctors' diagnoses was 2.0% of the population aged >15 years old (Kementerian Kesehatan, 2018), and is predicted to increase.

Carbohydrates increase blood sugar levels. Therefore, consuming large amounts of white rice should be avoided. The glycemic index is a value representing the relative ability of a carbohydrate food to increase the level of glucose in the blood. Foods with a low glycemic index minimize the rise in blood sugar levels and reduce the risk of diabetes. Colored rice varieties, such as those of red and black rice, have high nutritional value and contain anthocyanins that act as antioxidants and free radical scavengers to help prevent coronary heart disease (Xia et al., 2006). Small amounts of anthocyanin can prevent the production of bad fats—low density lipoproteins—and maintain or improve eyesight (Gunawan, 2005).

The Ministry of Health in the Republic of Indonesia reported that red rice contains 7.3% protein, 4.2% iron, and 0.34% vitamin B1 (Mukrie et al., 1995) as well as carbohydrates, fats, fiber, folic acid, magnesium, niacin, phosphorus, and vitamins A and C. Red rice flour can prevent various diseases, including colon cancer, kidney stones, constipation, hemorrhoids, high blood sugar and cholesterol levels (Suardi, 2005).

**Table 8.3** Chemical composition and bioactive compounds in two styles of red rice tea made from red rice grown in the Jatiluwih Village, Tabanan Regency, Bali (Budisanjaya et al., 2015)

| Compound | Oven-dried red rice tea | Roasted red rice tea |
|---|---|---|
| Water (g) | 5 ± 0.71 | 4 ± 0.06 |
| Ash (g) | 0.02 ± 0.008 | 0.02 ± 0.006 |
| Protein (g) | 8.22 ± 0.12 | 8.14 ± 0.06 |
| Fat (g) | 2.61 ± 0.21 | 2.53 ± 0.04 |
| Carbohydrate (g) | 81.41 ± 1.36 | 79.94 ± 0.03 |
| Anthocyanin (mg/100 g) | 0.25 ± 0.07 | 0.12 ± 004 |
| Phenol (mg GAE/L) | 18.56 ± 1.03 | 10.71 ± 0.78 |
| Tannin (%) | 1.33 ± 0.06 | 0.67 ± 0.05 |
| Antioxidant capacity (%) | 21.24 ± 0.12 | 14.59 ± 0.17 |

In the last few years, various processed products from red rice have been developed, including red rice tea that is in high demand by tourists who visit the Jatiluwih Village (Fig. 8.8). Processing red rice tea simply requires the red rice to be roasted in a cylindrical roaster that has been heated to 200 °C. The roaster is then continuously rotated manually for 7 min. The heat is then turned off and the roaster is rotated for a further 3 min. The red rice is then removed from the roaster and cooled (Budisanjaya, Wrasiati, & Wijaya, 2015). Alternatively, an oven can be used—unpolished red rice is placed in an aluminum pan and into a preheated oven (200 °C) for 5 min, before stirring and flipping with a wooden spoon, and heating for a further 5 min, before the oven is turned off and the pan is removed to allow the red rice to cool (Budisanjaya et al., 2015).

Using a roaster or an oven can result in differences in the chemical composition and bioactive compounds of the red rice tea (Table 8.3). Overall, the chemical compositions are similar, while the oven method produced higher levels of bioactive compounds and antioxidant capacity in the red rice tea than the roaster. However, roasted red rice tea had a higher tea extract content (31.23%) than oven-dried red rice tea (23.3%). The roasted red tea meets the minimum SNI tea extract content of 31% (Budisanjaya et al., 2015) (Fig. 8.8).

Red rice can also be processed into flour (Fig. 8.9), increasing the diversification of red rice, as it can be mixed with other ingredients or used as a substitute for wheat flour.

## 8.8 Research on Local Rice in Bali

### *8.8.1 Possible Genetic Erosion*

A study in 2015 found that the villages of Jatiluwih and Wongaya Gede cultivated two local white rice varieties (Mansur and Putih Cempaka), one red rice variety (Merah Cendana) and one black rice variety with black outer grain, but white inside

**Fig. 8.8** (**a**) Red rice plants; (**b**) Red rice; (**c**) Red rice tea; (**d**) Steeped red rice tea (Wijaya, Wrasiati, & Budisanjaya, 2014)

**Fig. 8.9** Red rice flour sold at the Tiara Dewata Super Market, Denpasar, Bali

(Jaka Selem) (Pharmawati, Wahyuni, & Wirasiti, 2015). However, by 2017, Putih Cempaka was rarely found in either village. According to local farmers, the high demand for red rice had resulted in farmers planting more local red rice than local white rice (Wayan Watra, pers. comm., Subak Keloncing, Wongaya Gede Village, Tabanan Regency, Bali, 2017).

### 8.8.2 Drought Stress Study

Pharmawati and Wrasiati (2018) studied the response of two Bali local rice cultivars to drought stress at the seedling stage. Root length and root to shoot ratios increased in both cultivars in the drought stress treatments induced by 20% and 30% PEG (polyethylene glycol). The two local Bali cultivars—Mansur and Putih Cempaka—had higher root to shoot ratios than IR64 in both the control and PEG treatments. Mansur and Putih Cempaka can be used as parent in developing new rice cultivars tolerant to abiotic stresses.

The same study examined the expression of the DREB (Dehydration-Responsive Element Binding) gene under drought, which increased in both Mansur and Putih Cempaka in response to 20% and 30% PEG treatments; in IR64, DREB expression only increased at 20% PEG and decreased at 30% PEG ((Pharmawati & Wrasiati, 2018). Several crops with better tolerance to drought, including rice, have increased DREB1 expression (Joshi et al., 2016).

### 8.8.3 Response of Rice Genotypes to Zinc Fertilizer Using RAPD

Global warming increase carbon dioxide ($CO_2$) levels in the atmosphere. Increased atmospheric $CO_2$ can reduce microelement contents, such as iron (Fe) and zinc (Zn) in some crop including rice cultivars tested (Zhu et al., 2018). Atomic absorption spectrometry did not detect any Zn in the local rice varieties from Jatiluwih Village—Mansur (white rice), Merah Cendana (red rice), Injin (black sticky rice), and Ketan Tahun (white sticky rice) (Defiani & Astarini, 2017).

As a result, Defiani, Astarini, and Pharmawati (2019) applied various levels of $Zn(SO_4).2H_2O$ fertilizer (0, 2.5, and 5.0 mg/kg soil) at 4 weeks after transplanting to Mansur, Merah Cendana, Injin, Ketan Tahun grown in polybags in an open field. At 8 weeks after transplantation, the application of $Zn(SO_4).2H_2O$ at 2.5 mg/kg soil increased plant height by 6% in white sticky rice, 12.6% in red rice, 17.5% in black sticky rice, and 16.5% in Mansur compared to control treatment (no $Zn(SO_4).2H_2O$ added. For Ketan Tahun (white sticky rice) and Mansur, Zn application increased the Zn concentration in the leaf blade of the last fully expanded leaf. The cultivars had different kinds of $Zn^{2+}$ absorption from soil solution to the leaf blades. The

**Table 8.4** Plant height (cm) of rice cultivars after application of various levels of $ZnSO_4.2H_2O$ fertilizer at 0, 4 and 8 weeks after transplanting (Defiani et al., 2019)

| Rice type | Plant age (week) | Plant height (cm) $ZnSO_4.2H_2O$ (mg/kg soil) | | |
|---|---|---|---|---|
| | | (0.0) | (2.5) | (5) |
| Sticky rice | 0 | 14.30 | 14.55 | 13.50 |
| | 4 | 23.83 | 36.67 | 34.17 |
| | 8 | 33.33 | 39.33 | 34.50 |
| Red rice | 0 | 17.17 | 16.92 | 16.92 |
| | 4 | 26.83 | 37.50 | 24.67 |
| | 8 | 32.33 | 45.00 | 26.00 |
| Black sticky rice | 0 | 14.33 | 18.58 | 19.00 |
| | 4 | 22.00 | 34.67 | 32.67 |
| | 8 | 28.00 | 45.50 | 44.33 |
| Rice cv. Mansur | 0 | 12.42 | 13.30 | 13.45 |
| | 4 | 23.33 | 41.83 | 28.67 |
| | 8 | 38.83 | 55.33 | 40.50 |

addition of 5 mg/kg soil of $ZnSO_4.2H_2O$ decreased Zn concentrations in the leaf blade of sticky rice, red rice, and Mansur (Table 8.4). A Random Amplified Polymorphic DNA (RAPD) marker was used to detect polymorphisms between treatments. Primers used were OPB12 and OPH1. Different PCR-RAPD band patterns were observed on Black Sticky Rice, Sticky Rice and Red Rice, shows that each genotype had a different response to Zn treatment.

### 8.8.4 *Mutation Breeding in Local Rice*

The National Nuclear Energy Agency of Indonesia (BATAN) has undertaken mutation breeding studies on several Bali local rice varieties. An 8-year study by Ita Dwi Mahyani on Merah Cendana found that irradiation using a gamma Cobalt-60 dose of 0.20 kg resulted in a shorter growing period and shorter plants. The new technology can reduce the height of Merah Cendana from 2 m to 1 m and the growing period from 6 months to 3 months. The mutant is also suitable for low land areas and maintains its quality. In addition, the leaves stay green even when the grain ripens. This was revealed when Ita Dwi Mahyani, researcher from BATAN harvested their first trial at Subak Pegedangan, Desa Pangkung Tibah in 2018. Farmers are excited about the research results, and hope that they will be able to grow local rice in their fields. Ita's long-term goal is to make this rice variety a national variety after trials in 16 areas in Indonesia (Balipost, 2018).

Ita Dwi Mahyani has also undertaken mutation breeding using gamma-ray irradiation on a local red rice variety (Cicih Gondrong) from Buleleng Regency in Bali, which has a long growing period and rice shoots/stalks lodge during maturing stage. Ita expects to be able to reduce the growing period, increase productivity, and produce stronger/sturdy stalks (Radar Bali, 2018).

## 8.9 Agrotourism Potential

Jatiluwih Village has the potential for agriculture-based tourism, based on the natural scenery of rice fields that extend for miles with breathtaking views of Mount Batukaru. Local people that maintain and preserve traditional rice farming system and planted local rice cultivars have become the main attraction for visitors from around the world.

Popular activities for tourists in Jatiluwih village include trekking, cycling, horse riding, local culinary (organic red rice are served in nearby restaurants), learning the philosophy of Subak and the local way of life, and visiting local temples and waterfalls near the village.

Facilities for visitors include:

1. Hotels and homestays
2. Restaurants and cafes
3. Parking areas
4. Attractive views for selfies and photography
5. Local guides for trekking
6. Waypoints and maps of trekking routes
7. Jatiluwih brochure
8. Tour of Jatiluwih village on VW Safari

Local communities and farmers are involved in Jatiluwih's growing tourism activities. Along the trekking routes, several stalls owned by locals sell coconut water, red, black, and white organic rice, traditional snacks, and souvenirs such as farmer's hats made from natural materials. The traditional 'rindik' instrument, made of bamboo, is usually played by farmers as a recreational activity in the rice terraces.

In December, farmers begin the process of planting traditional rice, starting with plowing the fields with the help of their cows or a plowing machine. During planting, tourists have the opportunity to transplant the available rice seedlings or feed the cows.

Tabanan Regency has interesting and attractive sites, including Tanah Lot, Jatiluwih, Bedugul Botanical Gardens, and Beratan Lake, which have the potential to increase the region's profit from tourism by as much as 40% from entry tickets and vehicles parking fee (Utama & Suyasa, 2018).

## 8.10 Conclusion

Local rice cultivation is an important activity for rural farming households in Jatiluwih village, and Bali in general. In addition to the farmer's daily needs, the production of local rice varieties is becoming important for income-generation for families.

## References

Badan Pusat Statistik (BPS). (2018). *Catalog no 5203023. Executive summary. 2018 harvested area and rice production in Indonesia*. BPS Jakarta, Indonesia, 17 pp.

Badan Pusat Statistik (BPS) Bali Province. (2017). *Area of land by Regency/Municipality and Type of use in Bali Province, 2017*.

Badan Pusat Statistik (BPS) Bali Province. (2019). *Berita resmi statistik pertumbuhan ekonomi Bali No. 14/02/51/Th. XIII*, 6 Februari 2019. Pertumbuhan ekonomi Bali tahun 2018, 12 pp.

Bahadoran, Z., Mirmiran, P., Delshad, H., & Azizi, F. (2014). White rice consumption is a risk factor for metabolic syndrome in Tehrani adults: A prospective approach in Tehran lipid and glucose study. *Archives of Iranian Medicine, 17*(6), 435–440.

Balipost. (2018). *Klon Beras Merah Cendana, G0 Direkomendasi Jadi Varietas Unggul Lokal*. http://www.balipost.com/news/2018/10/19/59063/Klon-Beras-Merah-Cendana,G0...html

BPTP. (2017a). *Mansur, padi lokal dataran tinggi*. http://bali.litbang.pertanian.go.id/ind/index.php/berita/51-info-aktual/738-mansur-padi-lokal-di-dataran-tinggi-bali

BPTP. (2017b). *Subak di Tabanan panen padi lokal*. http://bali.litbang.pertanian.go.id/ind/index.php/berita/51-info-aktual/810-subak-di-tabanan-panen-padi-lokal

BPTP Bali. (2019). *BPTP Bali Serahkan Sertifikat SDG ke Bupati Buleleng*. http://bali.litbang.pertanian.go.id/ind/index.php/berita/51-info-aktual/975-bptp-bali-serahkan-sertifikat-sdg-ke-bupati-buleleng. Accessed 1 Jan 2020.

Budisanjaya, I. P. G., Wrasiati, L. P., & Wijaya, I. M. A. S. (2015). Pengaruh Metode Pemanasan Terhadap Karakteristik Mutu Beras Merah Jatiluwih. *Media Ilmiah Teknologi Pangan, 2*(2), 115–121.

Budiwati, G. A., Kriswiyanti, E., & Astarini, I. A. (2019). Biological aspects and family relationship of local rice (*Oryza sativa* L.) in Wongaya Gede Village, Penebel, Tabanan Regency, Bali. *Metamorfosa, Journal of Biological Sciences, 6* (2), 277-292. https://doi.org/10.24843/metamorfosa.v06.i02.p20

Defiani, M. R., & Astarini, I. A. (2017). Zn and Fe micronutrients content in rice cultivar grown at Jatiluwih Village, Tabanan Regency, Bali. In *Proceeding of the international conference on biosciences*, 26–27 July 2016, Bali, Indonesia.

Defiani, M. R., Astarini, I. A., & Pharmawati, M. (2019). Response of rice genotypes to zinc fertilizer detected using RAPD. *IOP Conference Series: Earth and Environmental Science, 347*(1), 1–7. https://doi.org/10.1088/1755-1315/347/1/012088

Diamond, J., & Bellwood, P. (2003). Farmers and their languages: The first expansions. *Science, 300*, 597–603.

FAO. (2020). http://www.fao.org/plant-treaty/overview/en/. Accessed 19 Jan 2020.

Glaszmann, J. C. (1987). Isozymes and classification of Asian rice varieties. *Theoretical and Applied Genetic, 74*, 21–30.

Gunawan, A. (2005). *Anthocyanin menjaga kesehatan mata dan pembuluh darah*. Nirmala. November. pp.44.

Hasanah, I. (2007). *Bercocok Tanam Padi*. Azka Mulia Media. Jakarta, 68 pp.

Hu, E. A., Pan, A., Malik, V., & Sun, Q. (2012). White rice consumption and risk of type 2 diabetes: Meta-analysis and systematic review. *British Medical Journal, 344*, e1454.

Joshi, R., Wani, S. H., Singh, B., Bohra, A., Dar, Z. A., Lone, A. A., et al. (2016). Transcription factors and plants response to drought stress: Current understanding and future directions. *Frontier Plant Science, 7*, 1029–1029. https://doi.org/10.3389/fpls.2016.0102

Kementerian Kesehatan RI. (2018). *Hari diabetes Sedunia Tahun 2018*. Pusat Data dan Informasi, Kementerian Kesehatan RI.

Las, I., Suprihatno, B., Daradjat, A. A., Suwarno, B. A., & Satoto. (2004). Inovasi teknologi varietas unggul padi: perkembangan, arah, dan strategi ke depan. In F. Kasryno, E. Pasandaran, & dan A. M. Fagi (Eds.), *Ekonomi Padi dan Beras di Indonesia* (pp. 375–395). Jakarta, Indonesia: Badan Penelitian dan Pengembangan Pertanian.

Muhamad, K., Ebana, K., Fukuoka, S., & Okuno, K. (2017). Diversity and differentiation of *Oryza sativa* and *O. rufipogon* in Indonesia. *Genet Resources and Crop Evolution, 64*, 41–54. https://doi.org/10.1007/s10722-015-0331-6

Mukrie, N. A., Chatidjah, S., Masoara, S., Alhabsyi, A., Djasmidar, A. T., Bernadus, H. A., et al. (1995). *Daftar Komposisi Zat Gizi Pangan Indonesia*. Pusat penelitian Pengembangan Gizi, Direktorat Bina Gizi Masyarakat, Departemen Kesehatan Republik Indonesia.

Ngadi. (2013). Farmers empowerment in Subak Guama and Jatiluwih Tabanan Regency, Bali. *Pertanian, 11*(2), 149–162.

NusaBali.com. (2018). *Petani Bali kewalahan penuhi ekspor beras merah*. https://www.nusabali.com/berita/27469/petani-tabanan-kewalahan-penuhi-ekspor-beras-merah

Pharmawati, M., Wahyuni, I. G. A. S., & Wirasiti, N. N.. (2015). *Screening varietas padi lokal bali yang resisten terhadap kekeringan secara morpho-physiology dan analisis ekspresi gen*. Laporan Penelitian Fundamental. Universitas Udayana.

Pharmawati, M., & Wrasiati, L. P. (2018). Morpho-physiological and genetic responses of Bali local rice cultivars to drought stress at seedling stage. *Online Journal of Biological Sciences, 18*(1), 101–107.

Radar Bali. (2018). *Manfaatkan nuklir batan sukses persingkat masa panen padi beras merah*. https://radarbali.jawapos.com/read/2018/10/19/99319/manfaatkan-nuklir-batan-sukses-persingkat-masa-panen-padi-beras-merah

Silitonga, T. S. (2004). Pengelolaan dan Pemanfaatan Plasma Nutfah Padi di Indonesia. Balai Besar Penelitian dan Pengembangan Bioteknologi dan Sumberdaya Genetik Pertanian, Bogor. *Buletin Plasma Nutfah, 10*(2), 56–71.

Sitaresmi, T., Wening, R. H., Rakhmi, A. T., Yunani, N., & Susanto, U. (2013). Pemanfaatan Plasma Nutfah Padi Varietas Lokal dalam Perakitan Varietas Unggul. *Iptek Tanaman Pangan, 8*(1), 22–30.

Suardi, K. D. (2005). Padi beras merah: Pangan bergizi yang terabaikan? *Warta Litbang Pertanian, 27*(4), 1–2.

Subagyo, R. (2012). *Jatiluwih, menjaga warisan beras merah*. https://bali.antaranews.com/berita/32328/jatiluwih-menjaga-warisan-beras-merah. Cited 28 December 2019.

Suhartini, T. (2010). Keragamanan karakter morfologi plasma nutfah spesies padi liar (*Oryza* spp.). *Buletin Plama Nutfah, 16*(1), 17–28. https://doi.org/10.21082/blpn.v16n1.2010

Suharyanto, J. H., Mulyo, D. H. D., & Widodo, S. (2015). Analisis Produksi dan Efisiensi Pengelolaan Tanaman Terpadu Padi Sawah di Provinsi Bali. *Penelitian Pertanian Tanaman Pangan, 34*(2), 131–144.

Sutami, N. P., Londra, I. M., & Suastika, I. B. K.. (2016). *Upaya dan Kendala Penyediaan Benih Padi dari Pemerintah dan Petani Penangkar Mendukung Kedaulatan Pangan di Propinsi Bali*. Prosiding Seminar Nasional Inovasi Teknologi Pertanian. Banjarbaru, 20 Juli 2016, pp. 451–459.

Suwarno, E. L., & E. Soenarjo. (2001). Breeding of upland rice in Indonesia. In: M.K. Kardin, I. Prasadja, & dan M. Syam, (Eds.), *Upland Rice research in Indonesia current status and future direction* (pp. 1–6). Bogor, Indonesia: Central Research Institute for Food Crops.

Tanaka, K. (1998). Malayan cultivated rice and its expansion. *Agricultural Archaeology-Part IV, 1*, 344–350.

Undang-Undang Republik Indonesia Nomor 4 Tahun. (2006). *Tentang Pengesahan international treaty on plant genetic resources for food and agriculture* (Perjanjian Mengenai Sumber Daya Genetik Tanaman Untuk Pangan Dan Pertanian).

Utama, I. G. B. R., & Suyasa, N. L. C. P. S. (2018). *Segmentasi Pengunjung Daya Tarik Wisata: Warisan Budaya Dunia Jatiluwih*. Yogyakarta, Indonesia: Deepublish.

Wijaya, I. M. A. S., Wrasiati, L. P., & Budisanjaya, I. P. G.. (2014). *IbM Pengrajin The Beras Merah. Laporan Akhir Ipteks bagi masyarakat*. Fakultas Teknologi Pertanian. Universitas Udayana.

Xia, X., Ling, W., Ma, J., Xia, M., Hou, M., Wang, Q., et al. (2006). An anthocyanin-rich extract from black rice enhances atherosclerotic plaque stabilization in apolipoprotein E-deficient mice. *The Journal of Nutrition, 136*, 2220–2225.

Zhu, C., Kobayashi, K., Loladze, I., Zhu, J., Jiang, Q., Xu, X., et al. (2018). Carbon dioxide ($CO_2$) levels this century will alter the protein, micronutrients, and vitamin content of rice grains with potential health consequences for the poorest rice-dependent countries. *Sciences Advances, 4*(5), 1–9.

# Chapter 9
# Legislative Support for Agricultural Innovation in India

**Michael Blakeney**

**Abstract** The chapter looks at the role of intellectual property law in fostering agricultural innovation in India, particularly through patents and plant variety protection. Specifically, it surveys the Protection of Plant Varieties and Farmers' Rights Act, 2001 (PPVFR Act), the Seeds Act, 1966 and the Geographical Indications of Goods (Registration and Protection) Act, 1999 (GIs Act). A detailed examination is undertaken of the protection of farmers varieties under the PPVFR Act and of genetic resources under that Act. The legislative scheme of the GIs Act is detailed and its application to rice cultivation. The role of geographical indications in agricultural innovation is considered, as well as their relationship to traditional knowledge. The role of the Seeds Acts and Indian Seeds policies in promoting agricultural innovation is examined as well as the impact of the Biological Diversity Act 2002.

**Keywords** Protection of Plant Varieties and Farmers' Rights Act · 2001 (PPVFR Act) · Seeds Act · 1966 · Geographical Indications of Goods (Registration and Protection) Act · 1999 (GIs Act) · Biological Diversity Act 2002

## 9.1 Introduction

The Protection of Plant Varieties and Farmers' Rights Act, 2001 (PPVFR Act), the Seeds Act, 1966 and the Geographical Indications of Goods (Registration and Protection) Act, 1999 (GIs Act) were enacted to foster agricultural innovation in India. The PPVFR Act and the GIs Act are pieces of intellectual property legislation that were enacted to discharge India's obligation as a member of the World Trade Organization (WTO) to apply the provisions of the WTO Agreement on Trade Related Aspects of Intellectual Property Rights (TRIPS) (Singh & Aggarwal, 2013).

M. Blakeney (✉)
School of Law, The University of Western Australia, Crawley, WA, Australia
e-mail: michael.blakeney@uwa.edu.au

 
M. Blakeney, K. H. M. Siddique (eds.), *Local Knowledge, Intellectual Property and Agricultural Innovation*, https://doi.org/10.1007/978-981-15-4611-2_9

Article 27.3(b) of TRIPS obliges WTO Member States to "provide for the protection of plant varieties" and Article 22(2) of TRIPS obliges members to "provide the legal means for interested parties to prevent" the misleading use of geographical indications. The advantage for countries complying with the TRIPS Agreement is suggested in Article 7 of the TRIPS Agreement which states that the protection and enforcement of intellectual property rights "should contribute to the promotion of technological innovation and to the transfer and dissemination of technology, to the mutual advantage of producers and users of technological knowledge…." On 1 January 1995, India became a member of the World Trade Organization (WTO), which currently includes 164 member states (WTO, 2019).

It has been pointed out that compliance with TRIPS involves compliance costs, including the direct costs of ensuring that the country's legal, administrative and enforcement infrastructure can accommodate TRIPS implementation and indirect costs associated with more technologies being patented in response to TRIPS implementation and proprietors charging higher prices for access to their newly patented technologies (McCalman, 2001). A 2002 World Bank Study estimated that poor countries would have to pay an additional $US20 billion to foreign IPR rights holders as a result of TRIPS implementation (World Bank, 2002 and see also Maskus, 2000).

The payoff for countries complying with the TRIPS is suggested in Article 7, that the protection and enforcement of intellectual property rights "should contribute to the promotion of technological innovation and to the transfer and dissemination of technology, to the mutual advantage of producers and users of technological knowledge…." This chapter considers the extent to which the incentive thesis expressed in Article 7 of the TRIPS Agreement has underpinned agricultural innovation in India.

In India, agriculture provides the means for livelihood to more than 65% of the population and most of the farming population comprises small farmers (see Arjun, 2013). Consequently, legislative support for agricultural innovation is important and must take into account small and traditional farmers.

The Seeds Act, 1966 and Seeds Control Order enacted thereunder, and the New Policy on Seeds Development, 1988, are the basis for the promotion and regulation of the Indian seed industry. The New Policy on Seed Development had the aim of "providing to the farmer the best planting materials available in the world so as to increase productivity and thereby increasing farm income and export earnings" (Ministry of Agriculture, 1988). A new National Seeds Policy was promulgated in 2002 to account for the arrival of recombinant DNA technology, which created the possibility of patenting useful plant traits (see Blakeney, 2016a). The 2002 Policy stated its objective to create "a facilitative climate for growth of a competitive and localised seed industry" and encouragement of the importation of useful germplasm as "core elements of the agricultural strategy of the new millennium" (Ministry of Agriculture, 2002). The 2002 Policy sought to achieve its objectives on the twin pillars of plant variety rights legislation and an updated Seeds Act.

The promotion of agricultural innovation through the PPVFR Act, GIs Act and Seed Law has to also consider the Biological Diversity Act 2002, which aims the promote of the conservation, sustainable use and equitable sharing of benefits of India's biodiversity resources, including habitats, cultivars, domesticated stocks and breeds of animals and micro-organisms (Gadgil, 2003). There is an overlap of jurisdiction between the PPVFR Act and the Biological Diversity Act with respect to benefit-sharing arising from access to agricultural plant resources. Critically, the Biological Diversity Act provides that no person, whether Indian or foreign, shall apply for any intellectual property rights in or outside India for any invention based on research or information on biological resources obtained from India without the approval of the National Biodiversity Authority, established under the Act. The details of these four pieces of legislation are set out below.

## 9.2 Intellectual Property and Agriculture in India

The key provisions of the TRIPS Agreement that are relevant to agricultural innovation are Article 27.1 which requires that "patents shall be available for any inventions, whether products or processes, in all fields of technology…" and that "patents shall be available and patent rights enjoyable without discrimination as to … the field of technology…." Article 27.3(b) of the TRIPS Agreement requires WTO Members to "provide for the protection of plant varieties either by patents or by an effective sui generis system or by any combination thereof." Article 22.2 of the TRIPS agreement requires WTO Members to "provide the legal means for interested parties to prevent" the misleading use of geographical indications.

## 9.3 Patents

In conflict with the TRIPS patent obligation in Article 27.1 to not discriminate between fields of technology, India's Patents Act, 1970 in section 3(h) expressly excludes from patentability "a method of agriculture or horticulture". Thus, for example, a 2007 patent application for a method of reducing mycotoxin contamination of a plant or harvested material[1] was rejected by the Controller of Patents as not being an invention because it was an agricultural process. He noted that the application involved "agricultural techniques that are routinely used in agriculture by farmers for the plant protection by applying chemicals on the seed before sowing" (Sen, 2019).

Also excluded from patentability by s.3(j) are "plants and animals in whole or any part thereof other than micro-organisms but including seeds, varieties and species and following an amendment in 2002, essentially biological processes for production or propagation of plants and animals". This would appear to exclude the

[1] 9827/DELNP/2007.

patenting of plant breeding methods, although these have been patented in the USA (see Chap. 5).

The possibility of patenting genetically engineered plants was addressed in the 2013 Indian Intellectual Property Appellate Board case *Monsanto Technology LLC v Controller of Patents and Designs.*[2] Monsanto applied to patent a method for producing a transgenic plant that was capable of withstanding harsh environmental conditions. Monsanto had argued that by inserting a rDNA molecule into the plant, it had created a new invention. The Controller of Patents and Designs ruled that the application was disqualified by s.3j as being an essentially biological process. The Appellate Board ruled that it was insufficiently inventive.

A more recent decision has shed some light on the patentability of agri-biotechnological inventions. In *Monsanto Technology LLC v Nuziveedu Seeds Ltd,*[3] Monsanto challenged the April 2018 order of a two-judge bench of the Delhi High Court who revoked Monsanto's Indian patent (Indian patent number 214436) that covered a nucleic acid construct encoding a *Cry2b Bacillus thuringiensis* endotoxin protein (the *Bt-gene*) and a method for producing a transgenic plant with the *Bt-gene* with pest-resistance properties in several plants, including cotton. In its decision of 8 January 2019, the Indian Supreme Court sent the case back to a single judge of the Delhi High Court to decide whether invention met the requirements of the Patents Act.

A potential obstacle to the patenting of an invention comprising biological material is s.6(1) of the Biological Diversity Act 2002, which provides that "no person shall apply for any intellectual property right, by whatever name called, in or outside India for any invention based on any research or information on a biological resource obtained from India without obtaining the previous approval of the National Biodiversity Authority before making such application". Sub-section 6(2) provides that the National Biodiversity Authority may, while granting its approval "impose benefit-sharing fee or royalty or both or impose conditions including the sharing of financial benefits arising out of the commercial utilisation of such rights." These provisions are supplemented by an amendment to the Patents Act in 2005 that requires applicants for patents to "disclose the source and geographical origin of the biological material in the specification, when used in an invention."[4]

However, notwithstanding this seemingly harsh climate for agricultural patenting it has been reported that between 2013–2016, some 3000 patent applications were filed in India, including for herbicides, plant growth regulators, and processes for obtaining plant cells and plant tissue cultures (Sen, 2019). The fear has been expressed that "broad and strategic patenting by biotech companies may erect formidable entry barriers in biotechnology, promoting monopolistic control over the seed industry" (Pal, Tripp, & Louwaars, 2007). This is concerning, particularly as

---

[2] (IPAB) Order No. 146 of 2013.

[3] 2019 SCC OnLine SC 25.

[4] Patents Act 1970, s.10(4)(d)(ii)(D).

the public agricultural research sector is playing a less significant role in seed development in India (Srinivasan, 2004).

## 9.4 Plant Variety Rights

### *9.4.1 UPOV Convention*

Article 27.3(b) of TRIPS obliges WTO Member States to "provide for the protection of plant varieties". The history of legal protection for breeders of new plant varieties dates back to the promulgation of the International Convention for the Protection of New Plant Varieties (UPOV) (see Chap. 5). India is not a signatory to UPOV, although its membership has been mooted on a number of occasions (see Ranjan, 2009). Can a case be made for plant variety protection to act as an engine of agricultural innovation? A 2005 report of UPOV on the impact of plant variety protection in Argentina, China, Kenya, Poland and the Republic of Korea argued that UPOV membership increased breeding activities and the availability of new varieties developed at home or imported (UPOV, 2005) A contemporaneous World Bank study (Louwaars et al., 2005) observed that India had developed a vibrant seed breeding industry without intellectual property protection. It has been suggested that in countries such as India and China, where it is difficult to ensure the physical security of inbred lines due to proximity of plots with competing enterprises, plant variety rights protection is welcomed for protecting hybrid varieties (Correa, 2015). A 2004 survey of Indian seed breeders suggested that diversification of farmers into self/open pollinated varieties would be contingent upon the effective implementation of plant variety protection (Srinivasan, 2004). The survey also indicated that the lack of this protection was a major constraint for obtaining elite varieties from abroad.

### *9.4.2 The PPVFR Act*

India's implementation of its TRIPS obligation to protect plant varieties was the Protection of Plant Varieties and Farmers' Rights Act, 2001 (PPVFR Act). The PPVFR Act borrows elements from the UPOV Convention and contains elements, such as the protection of farmers' varieties, which are unique to India. However, it has been pointed out that enactment of the PPVFR Act was largely explained, not so much as a measure to encourage agricultural innovation, but to comply with India's TRIPS obligations (Kochupillai, 2011).

The objectives of the PPVFR Act, as enunciated in its preamble are to (i) recognize and protect the rights of farmers in respect of their contribution towards conserving, improving and making available plant genetic resources for the development of new plant varieties; (ii) protect plant breeders rights to accelerate agricultural development in the country; (iii) incentivise both the public and private sector to invest in R&D for the development of new plant varieties (especially those suited to Indian climatic and other conditions); (iv) facilitate the growth of the seed industry in India to ensure the availability of high quality seed and planting material to farmers; and (v) give effect to sub-paragraph (b) Article 27(3) of the TRIPs Agreement.

Protection under the PPVFR is afforded to a 'breeder' or persons claiming through the breeder who is defined in section 2(c) as "a person or group of persons or a farmer or group of farmers or any institution which has bred, evolved or developed any variety". The PPVFR encourages innovations in plant breeding, by establishing a system for the registration of plant varieties. Upon registration, a period of exclusivity is granted to the owners of registered varieties during which they can recoup their R&D costs and make a reasonable profit. Section 24(6) of the PVPFR Act provides that the periods during which the owners of a variety have the exclusive right to exploit and commercialise a registered variety are up to a total of: 18 years in the case of trees and vines and 15 years for any other plant variety.

To secure registration of a plant variety, section 15 of the PVPFR Act provides that it must be novel, distinct, uniform and stable. Each of these terms are defined in section 15. Thus 'novelty' is defined in section 15(3)(a) to mean that a variety has not previously been sold or disposed of by the breeder. Distinctness is defined in paragraph (b) to require that a variety "is clearly distinguishable by at least one essential characteristic from any another variety whose existence is a matter of common knowledge in any country at the time of filing of the application". Uniformity is defined paragraph (c) requiring that a variety, "subject to the variation that may be expected from the particular features of its propagation … is sufficiently uniform in its essential characteristics". Stability is defined in paragraph (d) to require that the "essential characteristics remain unchanged after repeated propagation or, in the case a particular cycle of propagation, at the end of each such cycle". Varieties which are "essentially derived" from a registered variety can also be registered under section 23 of the PPVFR Act, provided that they meet the criteria listed in section 15.

Whether the UPOV-style approach taken in the PPVFR accords with the science of plant breeding is increasingly being questioned. Janis and Smith note that it is outmoded to focus upon a phenotypic paradigm, based upon 'characteristics' and 'features' as plant breeding moves towards a genotypic approach, using genetic modification and molecular breeding techniques (Janis & Smith, 2007). In contrast, the new technologies are not a substitute for plant breeding but tools to supplement traditional methods (see Helfer, 2007; Sanderson, 2007; Sanderson & Adams, 2008).

## 9.5 Farmers' Rights

Standing at the threshold of much agricultural innovation are new varieties of landraces (traditional varieties) of crops cultivated by subsistence farmers. Traditional varieties account for around 60% of cultivated land and provide some 20% of the world's food (Wood & Lenne, 1997). From the early 1980s, civil society groups with an agricultural interest proposed the recognition of the contribution made by traditional farmers in conserving valuable biological resources (Andersen, 2016). This proposal was picked up by the Commission on Plant Genetic Resources of the Food an Agricultural Organization of the United Nations (FAO) under the chairmanship of Professor M.S. Swaminathan, which introduce the concept of farmers' rights in a voluntary International Undertaking on Plant Genetic Resources for Food and Agriculture (PGRFA) (Seema, 2012). This Undertaking was formalised in 2001 with the promulgation by the FAO of the International Treaty on PGRFA. The preamble to the Treaty acknowledged that PGFRA "are the raw material indispensable for crop genetic improvement" and affirmed "that the past, present and future contributions of farmers in all regions of the world, particularly those in centres of origin and diversity, in conserving, improving and making available these resources, is the basis of Farmers' Rights".

The Preamble explained that that "fundamental to the realization of Farmers' Rights, as well as the promotion of Farmers' Rights at national and international levels" were the rights "to save, use, exchange and sell farm-saved seed and other propagating material, and to participate in decision-making regarding, and in the fair and equitable sharing of the benefits arising from, the use of plant genetic resources for food and agriculture".

Article 9.2 of the Treaty envisaged that "the responsibility for realizing Farmers' Rights, as they related to Plant Genetic Resources for Food and Agriculture, rested with national governments" and that national legislation should include measures relating to:

(a) protection of traditional knowledge relevant to plant genetic resources for food and agriculture;
(b) the right to equitably participate in sharing benefits arising from the utilization of plant genetic resources for food and agriculture;
(c) the right to participate in making decisions, at the national level, on matters related to the conservation and sustainable use of plant genetic resources for food and agriculture.

The concept of Farmers' Rights was developed as "a counterbalance to intellectual property rights" (FAO, 1994). Farmers' rights were intended to promote a more equitable relationship between the providers and users of germplasm by creating a basis for farmers to share in the benefits derived from the germplasm that they had developed and conserved over time (See Glowka, 1998). Farmers' rights are conceived of as a 'retrospective equity' (Brush, 1996), primarily as the recognition of the moral obligation, rather than an economic incentive.

India became the first country to recognize farmers' rights in the PPVFR Act 2001.

## 9.6 Protection of Farmers' Varieties

The PPVFR, unique among national schemes for the protection of plant varieties, contains a scheme of protection for 'farmers' varieties'. Section 2(l) of the PPVFR Act defines as a 'farmers' variety' as a variety that—

(*i*) has been traditionally cultivated and evolved by the farmers in their fields; or
(*ii*) is a wild relative or land race of a variety about which the farmers possess the common knowledge

'Farmer' is defined in section 2(k) to mean any person who—

(*i*) cultivates crops by cultivating the land himself; or
(*ii*) cultivates crops by directly supervising the cultivation of land through any other person; or
(*iii*) conserves and preserves, severally or jointly, with any person any wild species or traditional varieties or adds value to such wild species or traditional varieties through selection and identification of their useful properties.

Section 39 of the PPVFR Act provides for the registration of farmers' varieties and s.24(1) provides for the issue of a certificate of registration. On receipt of a copy of the certificate of registration, s.24(1) provides that the Protection of Plant Varieties and Farmers' Rights Authority, established under the PPVFR Act, may invite claims of benefit-sharing in relation to the registered variety. This benefit sharing may relate both to farmers' varieties and new varieties that may be derived from them. In assessing claims, the Authority is required by s.26(5) to take into account: (*a*) the extent and nature of the use of genetic material of the claimant in the development of the variety relating to which the benefit-sharing has been claimed, and (*b*) the commercial utility and demand in the market of the variety relating to which the benefit-sharing has been claimed. Section 26(6) requires the amount of benefit sharing to be deposited by a breeder in the National Gene Fund, established under the PPVFR Act.

There is no evidence of any payments made by the National Gene Fund to farmers. In the first instance, the registration of farmers varieties has been quite low. For example, *The Plant Variety Rights Journal of India*, which is published by the Protection of Plant Varieties & Farmers' Rights Authority (PPVFRA), has records for the registration of 20 Farmers' varieties of rice cultivated in Kerala. A survey conducted by the Kerala Agricultural University during November 2018 identified 105 traditional varieties of rice in the region, of which 62 were being cultivated (KAU, 2018). Interestingly, all 20 of the Kerala rice varieties were registered by an agency (Seed Care) of the M. S. Swaminathan Research Foundation (MSSRF), a not-for-profit trust concerned with agricultural and rural development. The MSSRF has indicated that it will no longer be registering farmers' varieties as "the Biological Diversity Act 2002 gives protection to community rights if such varieties have been included in the Peoples Biodiversity Registers."[5] These registers have been created under s.22(6) of the Biological Diversity Rules.

A survey of registered farmers' varieties to 31 March 2016 lists the registration of three varieties each of wheat, sorghum and pigeon pea, five varieties of maize and 749 varieties of rice, of which 694 were filed by the State of Odisha (Das et al., 2019). Five rice varieties were file by farmers (PPVFRA, 2018). The low numbers of farmers' variety registrations has been attributed to a lack of knowledge of the

[5] Correspondence with the authors, 26 November 2018.

legislation, the shortage of resources and complexity of the registration process (Lushington, 2012).

## 9.7 Genetic Resources under the PPVFR Act

An important source of agricultural innovation are the genetic resources conserved by traditional farmers. Section 39(1)(iii) of the PPVFR Act provides that "a farmer who is engaged in the conservation of genetic resources of land races and wild relatives of economic plants and their improvement through selection and preservation" shall be entitled to recognition and reward from the National Gene Fund, established under section 45 of the Act. This is provided that conserved material has been used "as donors of genes" in varieties registrable under the Act.

Where a breeder or other person making application for registration of any variety under the Act makes use "of genetic material conserved by any tribal or rural families in the breeding or development of such variety", section 40 of the Act requires this to be disclosed in the application for registration.

Section 41 provides that a claim may be submitted to the National Gene Fund "on behalf of any village or local community in India" which has contributed to "the evolution of any variety". The section sets up machinery for the verification of such a claim and for the relevant breeder to pay the compensation into the National Gene Fund, which will then be paid to the claimants.

Thus far, in relation to the conservation of traditional rice varieties in Kerala, one farmer has received a "plant genome saviour community award" presented to him in 2016 by the Protection of Plant Varieties and Farmers' Rights Authority (Shaji, 2018), but otherwise there does not appear to have been any payments made to farmers from the National Gene Fund.

The recognition of the rights of farmers and communities in relation to the conservation of genetic resources is an aspect of the International Treaty on Plant Genetic Resources for Food and Agriculture, 2001 ("the Treaty"), which India ratified on 10 June 2002. Article 9.2 of the Treaty envisaged that "the responsibility for realizing Farmers' Rights, as they relate to Plant Genetic Resources for Food and Agriculture, rests with national governments" and that national legislation should include measures relating to:

(a) protection of traditional knowledge relevant to plant genetic resources for food and agriculture;
(b) the right to equitably participate in sharing benefits arising from the utilization of plant genetic resources for food and agriculture;
(c) the right to participate in making decisions, at the national level, on matters related to the conservation and sustainable use of plant genetic resources for food and agriculture.

The commercial value of genetic resources conserved by farmers in developing new varieties is difficult to quantify. The value of farmers' varieties does not directly depend on their current use in conventional breeding, due to the modest gene flow from landraces to privately marketed cultivars of major crops because conventional breeding has focused on crosses among elite materials from breeders' own collections and advanced lines developed in public institutions (Wright, 1998). In contrast, about 6.5% of all genetic research undertaken in agriculture has focussed on germplasm derived from wild species and land races (McNeely, 2001). Certainly, in this time of climate change, breeders will increasingly resort to traditional varieties that can withstand agricultural stresses, such as increases in temperature, fluctuations in rainfall and pests and moulds.

Geographical Indications of Goods (Registration and Protection) Act, 1999 (GIs Act).

## 9.8 Legislative Scheme

The GIs Act, enacted on 30 December 1999, did not come into force until 15 September 2003. This Act does not contain a preamble stating its objectives, other than "to provide for the registration and better protection of geographical indications relating to goods." The Act is administered by the Controller General of Patents, Designs and Trade Marks, who is the Registrar of Geographical Indications with the Geographical Indications Registry located in Chennai.

The definition of geographical indications (GIs) in section 2(1) of the GI Act utilises the language of TRIPS Article 22.1, requiring an association between the quality or characteristics of goods and their place of production. It states that

> "geographical indication", in relation to goods, means an indication which identifies such goods as agricultural goods, natural goods or manufactured goods as originating, or manufactured in the territory of a country, or a region or locality in that territory, where a given quality, reputation or other characteristic of such goods is essentially attributable to its geographical origin and in case where such goods are manufactured goods one of the activities of either the production or of processing or preparation of the goods concerned takes place in such territory, region or locality, as the case may be.

An explanation appended to this provision states that "for the purposes of this clause, any name which is not the name of a country, region or locality of that country shall also be considered as the geographical indication if it relates to a specific geographical area and is used upon or in relation to particular goods originating from that country, region or locality as the case may be". This explanation was probably inserted to deal with the protection of the GI "Basmati". There is no geographical location which has that name, but its use has been associated with rice production in India. "Basmati Rice" was registered as a geographical indication for rice produced in the states of Punjab, Haryana, Delhi, Himachal Pradesh, Uttarakhand, and parts of western Uttar Pradesh and Jammu & Kashmir.[6] Madhya Pradesh, Rajasthan and Bihar were excluded from this registration, on March 15, 2018, as not being in the traditional Basmati rice growing area in the Indo-Gangetic Plain (Rana & Co, 2018).

[6] Reg. No 145.

Goods are defined in section 2(1) to mean "any agricultural, natural or manufactured goods or any goods of handicraft or of industry and includes food stuff".

An indication is defined to include "any name, geographical or figurative representation or any combination of them conveying or suggesting the geographical origin of goods to which it applies". This is important in places where literacy might be low, and the geographical origin is indicated by symbols representing the place of production.

The GIs Act establishes a system for the registration of GIs. Section 6 requires a "Register of geographical indications" to be kept at the head office of the Geographical Indications Registry in Chennai in which shall be entered all registered geographical indications with the names, addresses and descriptions of the proprietors, the names, addresses and descriptions of authorised users".

Excluded from registration by s.9 of the Act are false, confusing, misleading or deceptive geographical indications or those which comprises or contains scandalous or obscene matter, or any matter likely to hurt the religious susceptibilities of any class or section of the citizens of India, or which are determined to be generic, which have ceased to be protected in their country of origin, or which have fallen into disuse in that country. For example, the Cour d'appel d'Orléans ruled in 1926 that 'Camembert' had become a generic description of a type of soft cheese (noted in Gangjee, 2016).

Section 11(1) of the GIs Act provides geographical indications may be registered by "any association of persons or producers or any organization or authority established by or under any law for the time being in force representing the interest of the producers of the concerned goods…" In the early years of the GIs Act, applicants for foodstuff GIs tended to be government departments or statutory boards or enterprises, or university agriculture departments who were not directly involved in the production of the goods. This may well have changed with the refusal in 2009 of the registration of a GI for Ganjam Goat Ghee by Orissa Veterinary College on the grounds that it could not adequately demonstrate that it represented the interest(s) of the producers (Vinayan, 2017).

Acceptable applicants are those organizations involved in ensuring that farmers cultivating the varieties embraced by GI registrations, adhere to prescribe cultivation and processing standards. This standard-monitoring activity has the effect of preserving the commercial reputation of the GI. For example, producers of 'Darjeeling Tea' can only use the geographical indication if they produce their tea according to the production standards that have been prescribed for the GI (Chaudhary, 2019).

Section 14 provides for oppositions to be taken to applications for registration. The usual ground of opposition is that there is particular linkage between the designation and the quality or character of a product. An example of a successful opposition to the registration in 2013 of 'Kalanamak' by an NGO as a GI for rice coming from eastern Uttar Pradesh. The opponents pointed out that both the description of the Kalanamak variety and its morpho-agronomic characteristics were incorrect and the places to benefit from the GI were haphazard (Chaudhary, 2019). A GI was subsequently granted for the rice, based on a more coherent agro-climatic zone (Chaudhary, 2019).

Section 18 provides that the duration of a GI is for 10 years and subject to the payment of a renewal fee, which may be renewed for additional 10-year period.

Section 21 permits the authorised user the exclusive right to the use of the GI in relation to the goods in respect of which the GI is registered, and the registered proprietor of the GI and the authorised user have the right to obtain relief, such as damages in respect of infringement of the GI.

Infringement is defined by s.22 as the unauthorised use of the GI by a person who indicates or suggests that their goods originate in a geographical area other than the true place of origin of such goods in a manner which misleads the persons as to their geographical origin or use which is an "act of unfair competition", meaning any "act of competition contrary to honest practices in industrial or commercial matters."

Criminal penalties are imposed by ss.39–44 in relation to the false use of a GI.

The first registered GI in India was for Darjeeling Tea in 2004; since then a number of GIs for agricultural products have been registered (Chaudhary, 2019; Dattawadkar & Mohan, 2012; Kumar & Srivastava, 2017). However, it has been noted that the registration of GIs in India for agricultural products has been hampered by a general lack of awareness of the GI system among farmers (Blakeney, Krishnankutty, Raju, & Siddique, 2019; Nanda, 2013; Vinayan, 2017).

## 9.9 GIs and Agricultural Innovation

GIs are particularly advantageous for the producers of agricultural products in allowing them to differentiate their products from general commodity products such as rice, coffee and tea, thereby enhancing market access and attracting premium prices (see Diallo, 2017). The principal reasons that have been identified for GI-marked goods attracting premium prices, are that consumers prize their exoticism (Agarwal & Barone, 2005) and the greater care in their production compared with undifferentiated commodity products (Réviron et al., 2009). Another factor, is the increasing realisation that traditionally produced goods are often freer from contaminants, such as herbicides and pesticides and that the GIs applied to these goods provides confidence in their traceability (Blakeney, 2017).

GIs can play an important role in signalling to consumers the quality of goods (Becker, 2008; Hobbs, 2003; Hobbs & Kerr, 2006). They are important for signalling credence attributes, particularly as an origin brand will be underpinned by a registration and certification system. These will be administered by a producers' association, which will secure compliance with agreed production standards. Producers can thus signal quality and the associated reputation that has been developed over time (Winfree & McCluskey, 2005) and incentivised by the premium prices attracted by a GI to maintain product quality (Moschini, Menapace, & Pick, 2008).

Of course, for the perceived benefits of GI labelling to be realised, such as the promotion of environmental sustainability, consumer awareness that origin labelling represents qualities linked to natural and human factors is needed. This ties in with the consumer demand for traceability in agrifood products (Murdoch, Marsden, & Banks, 2000; van der Ploeg, Renting, & Minderhoud-Jones, 2000). Rural product certification schemes have proliferated since the mid-1990s. They include the certification of organic agriculture, fair-trade certification of products from developing countries, and food produced in compliance with sanitary and traceability protocols (Giraud & Amblard, 2003: Mutersbaugh et al., 2005). Consumers have been identified as placing increasing value on the integrity of food, such as the social and environmental standards involved in the production and processing of agrifood products (Hobbs et al., 2005; Renting, Marsden, & Banks, 2003). This is particularly the case following recent food safety crises. As it is not unusual for food to be grown, processed and packaged in different places, consumer trust in products is eroded, particularly as a consequence of these crises. Studies indicate a willingness of consumers to pay a premium price to producers who offer transparency in relation to the composition and origin of their products. In situations where uncertainty about quality or safety is elevated, such as in a health crisis, origin labelling can become an important means of inferring product quality, e.g. meat labels after the BSE crisis in Europe (Becker, 2009; Lees, 2003; Loureiro & Umberger, 2007; Verbeke & Viaene, 1999) and dairy product labels after the Chinese Melamin crisis (Wu & Zhang, 2013.

Concerns about the safety of agrifoods in China has stimulated an interest in the mechanisms for assuring traceability in food chains. In this context GIs "may convey assumed 'local' (traceability) and 'natural' (nutritiousness and safety) characteristics thereby acting as proxies for quality" (Zhao, Finlay, & Kneafsey, 2014). In Europe, where GIs have been longest developed, there are some empirically based suggestions that consumers' and producers have expectations of the quality of origin products in the European market (Stasi et al., 2011; Teuber, 2011). It has been suggested that the EU ban on the importation of Alphonso mangos from Maharashtra, Goa, Karnataka and Gujarat could be overcome by the development of a GI for mango were the product specifications include sanitary and phytosanitary monitoring (Pai & Singla, 2016). Additionally, producers can formulate their product specifications by taking into account the positive environmental impacts of food cultivation (Belletti et al., 2015).

One of the justifications advanced for the establishment of an early GIs system for the protection of wines produced in France was the role that they played in preserving agriculture and rural employment in areas that were unsuitable for cereals and other crops (Stanziani, 2004). The protection of GIs in the EU accords with the its policy on rural development (see Blakeney, 2019). Recital 4 to Regulation (EU) No 1151/2012 of the European Parliament and the Council of 21 November 2012 on quality schemes for agricultural products and foodstuffs that govern GIs identifies that:

> Operating quality schemes for producers which reward them for their efforts to produce a diverse range of quality products can benefit the rural economy. This is particularly the case in less favoured areas, in mountain areas and in the most remote regions, where the farming sector accounts for a significant part of the economy and production costs are high. In this way quality schemes are able to contribute to and complement rural development policy …. In particular, they may contribute to areas in which the farming sector is of greater economic importance and, especially, to disadvantaged areas.[7]

The creation of local jobs through the protection of GIs is a factor that has been identified as retarding rural exodus (O'Connor & Co, 2005) For example, in employment has increased in the French Comté cheese industry, as opposed to areas that produce alternative generic cheeses (Gerz & Dupont, 2006). Barjolle (2016 identifies 21 European GIs where the maintenance of rural development is in the product specification. In 2018 the African Union (AU) formulated a Continental Strategy for Geographic Indications (GIs) in Africa, 2018–2023 "to facilitate sustainable rural development in line with the vision of African leaders of a prosperous Africa based on inclusive growth and sustainable development" (African Union, 2018). The AU envisaged that GIs for food and non-food products

> represent an answer to enhance exchanges among stakeholders at infra-national levels and thus to preserve and promote traditional products on local markets, as well as to position African export products better on international markets. In African countries, GIs can be used as a tool for the organization and promotion of agricultural value chains. They can create incomes for farmers and other stakeholders in the value chain, such as small processing units and petty traders, and therefore help them to face food lean periods and food and nutrition insecurity.

Considerable work has been done in Africa to identify agricultural products that could benefit from GIs protection (see Blakeney, et al., 2012), including Burundi tea and coffee, Gambian cashews, Ugandan cotton and vanilla, shea butter from Burkina Faso, shallots from the Dogon area of Mali, rooibos tea from South Africa, Galmi onions from Niger, Fouta Djalon potatoes from Guinea and Madagascar Vanilla (see Mengistie & Blakeney, 2016). Already registered as GIs in Morocco—the most advanced African country in this regard—are Argane (oil), Clementine of Berkane, Majhoul Dates of Tafilalet, Pomegranate Sefri Ouled Abdellah, Prickly Pear of Aît Baâmrane, Chefch Aouen's Goat Cheese, Aziza Bouzid Dates of Figuig, Uphorbia Honey of Tadla-Azilal, Almonds of Tafraout, Boufeggous Dates, Midelt Apple, Medlars of Zegzel, Arbutus Honey of Jbal My Abdess Alam, Keskes Khoumassi, or Keskes Moukhamess, Extra Virgin Oil of Ouezzane, Safi Capers, Jihel Dates of Drâa, Azilal Walnut, Eastern Rosemary Dried Leaves, Eastern Rosemary Essential Oils, Doukkali Raisin, Rif Almonds, Ait Ouabelli Henna, Oued El Maleh Quince, Outat El Haj Olive Oil, Nabout Dry Fig of Taounate, Tafersite Olive Oil, Honey of Desert Euphorbia, Tyout Chiadma Olive Oil, Saffron of Taliouine, Rose of Kelâat M'gouna-Dadès, Extra Virgin Olive Oil Aghmat Aylane and Oulmes Lavender Essential Oils. A tangible consequence of Morocco adopting a GIs law on the EU model is that ended an agreement with the EU in January 2015 for the reciprocal

[7] Official Journal L 343, 14.12.2012, p. 1.

protection of Moroccan and EU GIs for the mutual protection of their GIs for agricultural products and foodstuffs.[8]

In India, the Darjeeling tea GI has been identified as a conspicuous success, particularly because the quality control of tea is secured by the Tea (Marketing and Distribution Control) Order of 2000, read in conjunction with the Tea Act 1953.

## 9.10 GIs and Traditional Knowledge

The role of the traditional knowledge of indigenous and traditional peoples and traditional communities in identifying useful plants and germplasm is well-recognized (Blakeney, 2001, 2002). Proposals for an international convention to confer IP protection on traditional knowledge date back to 2000. Some 20 years later the negotiations at the World Intellectual Property Organization (WIPO) for a treaty on traditional knowledge are on-going, largely due to the inability of developed and developing countries to reconcile their positions on the subject (Blakeney, 2016b). In the absence of a *sui generis* piece of legislation to protect traditional knowledge, it has been suggested that TK legislation is the second-best alternative (Dagne, 2010;. A more optimistic assessment of the potential for GIs to protect TK was made by Panizzon and Cottier (2005) who observed that:

> Traditional Knowledge (TK) and Geographical Indications (GIs) share a common element insofar as they both protect accumulated knowledge typical to a specific locality. While TK expresses the local traditions of knowledge, GIs stand for specific geographical origin of a typical product or production method. GIs and TK relate a product (GIs) … [or] a piece of information (TK) [respectively] to a geographically confined people or a particular region or locality.

Similarly, in its Review of Existing Intellectual Property Protection of Traditional Knowledge the Secretariat of the Intergovernmental Committee (IGC), established by WIPO to administer negotiations for the traditional knowledge treaty, observed that:

> Geographical Indications as defined by Article 22.1 of the TRIPS Agreement and appellations of origin, as defined by Article 2 of the Lisbon Agreement … rely not only on their geographical connotation but also essentially, on human and/or natural factors (which may have generated a given quality, reputation or other characteristic of the good). In practice, human and/or natural factors are the result of traditional, standard techniques which local communities have developed and incorporated into production. Goods designated and differentiated by geographical indications, be they wines, spirits, cheese, handicrafts, watches, silverware and others, are as much expressions of local cultural and community identification as other elements of traditional knowledge can be.[9]

---

[8] See http://europa.eu/rapid/press-release_IP-15-3440_en.htm, accessed 4 December 2019.

[9] WIPO/GRTKF/IC/3/7, 6 May 2002,

Three examples provided by the IGC Secretariat of TK protected by geographical indications are: 'Cocuy Pecaya' liquor (from Venezuela), and 'Phu Quoc' fish sauce and 'Shan Tuyet Moc Chau' tea (both from Vietnam).[10]

Downes (2000) points out that GIs are

> especially suitable for use by indigenous and local communities since they are based upon collective traditions and a collective decision-making process; they protect and reward traditions while allowing evolution; they emphasize the relationships between human cultures and their local land and environment; they are not freely transferable from one owner to another; and they can be maintained as long as the collective tradition is maintained.

GIs reward the goodwill and reputation of producers who use traditional methods created or built up in a geographical territory (Cottier & Panizzon, 2004). In this way GIs protection can protect the traditional knowledge of local communities that have developed folk varieties from land races (Downes & Laird, 1999). GIs reward goodwill and reputation created over many years, while allowing evolution, making them suitable for the protection of traditional knowledge (Dagne, 2010).

The particular utility of GIs protection in the absence of a legal regime that protects traditional knowledge is that it recognizes the quality and reputation of the agricultural products of traditional communities and prohibits others from free-riding off the reputation of those products, as long as natural and cultural characteristics in the relevant place of cultivation are maintained (Blakeney, 2009; Cullet et al., 2006).

In several traditional communities, cultivated crops that may be both sources of food and medicine can also be the repository of religious and cultural traditions. Thus, for example, Navara rice from Kerala, registered under a GI, has medical properties, described as part of Ayurvedic treatment in the fifteenth century in India (Jagdish, Makanur, & Eraya, 2006). Two 2019 registrations of foods with cultural applications are: Palani Panchamirtham, from Palani Town in the Dindigul District of Tamil Nadu and Tirur betel vine from Kerala (The Hindu, 2019). The prasadam is made up of, banana, jaggery sugar, cow ghee, honey and cardamom in defined proportions. It is one of the main offerings in the abishegam of Lord Dhandayuthapani Swamy, a temple situated in Palani Town (Kandavel, 2019) Tirur betel vine from the Malappuram District, Kerala, is valued both for its mild stimulant action and medicinal properties and has cultural uses.

## 9.11 The Seeds Act 1966

"India has one of the most dynamic and diversified seed industries in the developing world" (Pal et al., 2007), which is attributed both to strong public research and supportive government policies providing for open access to publicly-bred germplasm

[10] Ibid., para. 13.

and fiscal incentives for investment in plant breeding. By 2003, India had more than 150 private seed companies along with 13 state seed corporations (Gadwal, 2003).

The Seeds Act, 1966 and Seeds Control Order enacted thereunder, and the New Policy on Seeds Development, 1988, were the basis for the promotion and regulation of the Indian seed industry. The New Policy on Seed Development had the objective of "providing to the farmer the best planting materials available in the world so as to increase productivity and thereby increasing farm income and export earnings" (Ministry of Agriculture, 1988). They were perceived to have made a significant contribution to the Green Revolution in India (Bhalla & Singh, 2001; Chakravarti, 1973; Parayil, 1992). A new National Seeds Policy was promulgated in 2002 to account for the arrival of recombinant DNA technology, which created the possibility of patenting useful plant traits. The 2002 Policy stated its objective of creating "a facilitative climate for growth of a competitive and localised seed industry" and encouraging importation of useful germplasm as "core elements of the agricultural strategy of the new millennium" (Ministry of Agriculture, 2002). The 2002 Policy sought to achieve its objectives on the twin pillars of plant variety rights legislation and an updated Seeds Act.

Clause 2.11 of the 2002 Policy provided that "seed exchange among farmers and seed producers will be encouraged to popularise new/non-traditional varieties" and clause 2.12 directed that "seeds of newly developed varieties must be made available to farmers with minimum time gap". To implement the new policy a Seeds Bill was introduced in the Rajya Sabha on 9 December, 2004 to replace the 1966 Act. The Bill met with opposition from farmers concerned about their traditional rights to seeds, as well as civil society and politicians concerned about the influence of foreign multinational seed companies and the threatened loss of biodiversity from monocultures. Responding to this criticism, the Seeds Bill 2004 has undergone three revisions. The most recent version, prepared in 2011 is pending in Parliament.

## 9.12 The Biological Diversity Act 2002

### *9.12.1 Legislative Scheme*

The key to the development of agricultural crops in India has been access by farmers and plant breeders to the country's considerable biological resources and associated traditional knowledge. Access to these resources and knowledge is regulated by the Biological Diversity Act 2002 (the Act) which was enacted by the Indian Parliament on 5 February 2003 to implement and give effect to the Convention on Biological Diversity (CBD). The Act was passed pursuant to Article 253 of the Indian Constitution, which empowers the Government of India to implement its international obligations through national legislation. The Act was also seen as a response to the furore surrounding the patenting of neem, basmati and turmeric by foreign firms (Sagar, 2005). Section 3 of the Act provides that no person, whether Indian or foreign can "obtain any biological resource occurring in India or knowledge associated thereto for research or for commercial utilisation or for bio-survey and bio-utilisation" without the approval of the National Biodiversity Authority (NBA), as established under the legislation. 'Biological resources' is defined in s.2(c) to include "plants, animals and micro-organisms or part thereof, their genetic material and by-products (excluding value added products) with actual or potential use or value". 'Commercial utilization' is defined in s.2(f) to include "means end uses of biological resources for commercial utilization such as … genes used for improving crops and livestock through genetic intervention" but does not include "conventional breeding or traditional practices in use in any agriculture, horticulture, poultry, dairy farming, animal husbandry or bee keeping". 'Bio-survey and bio-utilisation' is defined in s.2(g) to mean "survey or collection of species, sub-species, genes, components and extracts of biological resource for any purpose and includes characterisation, inventorisation and bioassay".

Section 6 of the Act provides that no application for IP rights may be made without the approval of the NBA "in or outside India" for any invention based on any research or information on a biological resource obtained from India, but excludes applications made under the PPFVRA.

Applications to the NBA for approval are made under s.19 of the Act and s.20 requires the approved person to obtain permission from the NBA for the transfer of any biological resource or associated knowledge. Section 21(1) of the Act states that the NBA while granting approvals under s.19 and s 20 shall ensure that there will be an "equitable sharing of benefits" arising out of the use of "accessed biological resources, their by-products, innovations and practices associated with their use and applications and knowledge relating thereto" in accordance with mutually agreed terms and conditions between the person applying for such approval, local bodies concerned and the benefit claimers. Section 2(a) defines 'benefit claimers' as "the conservers of biological resources, their by-products, creators and holders of knowledge and information relating to the use of such biological resources, innovations and practices associated with such use and application". Section 21(2) requires the NBA to evaluate the benefit-sharing arrangements by reference, *inter alia*, to the ownership of IP rights and payments to individuals or groups of individuals who provided biological resources or knowledge.

Section 36 of the Act requires the Central Government to develop national strategies, plans and programmes for the conservation and promotion and sustainable use of biological diversity and that it shall "endeavour to respect and protect the knowledge of local people relating to biological diversity" in line with recommendations of the NBA. Section 41 of the Act requires that every local body shall constitute a Biodiversity Management Committee within its area for the purpose of promoting conservation, sustainable use and documentation of biological diversity "including preservation of habitats, conservation of land races, folk varieties and cultivars, domesticated stocks and breeds of animals and micro-organisms and chronicling of knowledge relating to biological diversity". The explanatory notes to the section define 'cultivar' to mean "a variety of plant that has originated and persisted under cultivation or was specifically bred for the purpose of cultivation" a 'folk variety' means "a cultivated variety of plant that was developed, grown and exchanged informally among farmers"; and 'landrace' is defined as a "primitive cultivar that was grown by ancient farmers and their successors."

On 15 April 2004 the Biodiversity Rules (the Rules) were promulgated by the Indian Parliament to carry out the purposes of the Act. Rule 14 detailed the procedures for seeking access to biological resources and required the NBA to detail any restrictions on the transfer of accessed biological resources and traditional knowledge to any third party without prior approval. Most relevant to agricultural innovation was the requirement in Rule 22 that every local body shall constitute a Biodiversity Management Committee (BMC) within its area of jurisdiction. The main function of the BMC, according to Rule 22(6), is to prepare a "People's Biodiversity Register in consultation with local people" which shall contain comprehensive information on "availability and knowledge of local biological resources, their medicinal or any other use or any other traditional knowledge associated with them." In the first 10 years of the operation of the Rules 33,077 BMCs were established across 23 states of India, of which 27,712 were in Madhya Pradesh (Bhutani & Kohli, 2012). There is no legal protection available for the knowledge recorded in the register and no requirement that consent of local communities be sought in accessing the register (Kumar & Srivastava, 2019).

Arguably, the requirements of the Indian biological diversity regime are not in conflict with its IP regime. It should be noted in this regard that Art.16(5) of the CBD recognizes that patents and other IP rights may have an influence on the implementation of the Convention, but requires that "subject to national legislation and international law" signatories shall "ensure that such rights are supportive of and do not run counter to its objectives."

### *9.12.2 Litigation*

The first case brought by the NBA was its 2012 action against Monsanto, its Indian partner Maharashtra Hybrid Seeds Company, (Mahyco)—26% of which is owned by Monsanto— and its Indian collaborators, the University of Agriculture Sciences (UAS) at Dharwad in north Karnataka (UAS-Dharwad) and Sathguru Management Consultants Ltd., a private Indian company acting as a coordinator on behalf of USAID and Cornell University. The NBA decision charged these entities with alleged violation of the Act "for accessing and using the local brinjal (eggplant) varieties for development.

of *Bt* brinjal without prior approval of the competent authorities" (see Abdelgawad, 2012). *Bt* Brinjal, India's first GM food crop, was developed by inserting a crystal protein gene, developed by Monsanto from the soil bacterium, *Bacillus thuringiensis*, into the genome of various local eggplant cultivars to develop resistance to insect pests. Brinjal is prone to attack from insect pests and diseases, the most serious and destructive of which is the fruit and shoot borer (FSB) *Leucinodes orbonalis*. FSB larvae bore into tender shoots and fruits, retarding plant growth, making the fruits unsuitable for market and unfit for human consumption. Fruit damage as high as 95% and losses of up to 70% in commercial plantings have been reported (ISAAA, 2019). In 2006 the Indian Genetic Engineering Approval Committee (GEAC) set up an Expert Committee to look into *Bt* Brinjal; in October 2009, it declared *Bt* Brinjal safe and recommended its commercial approval to the environmental ministry which subsequently imposed a moratorium on the commercial release of the crop (Kumar, 2011).

The NBA action originated with a complaint made by the Environment Support Group (ESG), an NGO based in Bangalore, to the Karnataka Biodiversity Board in 2010. On 28 May 2011 the Karnataka Biodiversity Board informed the NBA that, 'six local varieties for development of *Bt* Brinjal were accessed in the state by the two companies without prior approval from the State Biodiversity Board and the NBA and called for legal action. The complaint by ESG was preceded by farmer protests and the announcement of a moratorium on *Bt* Brinjal by the Minister of Environment a public consensus on health and safety issues had been reached (Jebaraj, 2011).

Relevant to the question of agricultural innovation in India was that the complaint concerned an agreement between Mahyco, UAS-Dharwad, and Sathguru, which had the objective of developing 'pro-poor varieties of insect tolerant Bt. Eggplant' (Abdelgawad, 2012). Pursuant to this agreement, Mahyco transferred the Cry1AC gene technology (supplied by Monsanto) to the six local varieties provided by UAS-Dharwad, and the technology was transferred by Mahyco to UAS (and also to the Tamil Nadu Agriculture University) as a royalty-free license to make it available "to resource-constrained farmers" under a joint research project (Abdelgawad, 2012).

The NBA ruled that the research project seemed *prima facie* to fall outside the scope of guidelines issued by the Central Government and that the three parties should have obtained NBA approval. The complaint that prior notice had not been given to the Karnataka Biodiversity Board, as required by s.7 of the Act was dismissed as that provision concerned 'commercial utilization' and the joint research agreement concerned the use of *Bt* technology, to develop or distribute brinjal to resource-constrained farmers 'other than by sale'.

In October 2013, the Karnataka High Court dismissed pleas to halt criminal prosecution against senior representatives of the research partners (Sreeja, 2013). On 3 January, 2015, two day before the hearing of the case, the Registrar, Vice Chancellor and former Vice-Chancellor of University of Agricultural Sciences, Dharwad, obtained a 6-month stay of prosecution by the Dharwad Bench of the Karnataka High Court (Sood, 2015a), but this was vacated by the Karnataka High Court in August 2015 (Sood, 2015b). The matter is still pending (Veena & Rajasekharan, 2019). However, on 12 May 2019, Prashant Bhushan, a public interest lawyer, issued a legal notice in a letter to the Minister for Environment, Forest and Climate Change that the moratorium on the commercial cultivation of *Bt* Brinjal was being violated by a farmer in Karnataka (Todhunter, 2019). The letter is to be distributed to the Prime Minister, the Minister of Agriculture and all members of parliament, which may bring some political and policy resolution to the *Bt* Brinjal affair.

Interestingly, the *Bt* Brinjal controversy prompted the Mattu Gulla Growers Association, a traditional brinjal grower community, to protect its traditional brinjal variety, called Udupi Mattu Gullazz, through its registration under the GIs Act in May 2011. This registration denoted its origin in Mattu Village, Udupi (GIs Registry, 2011).

## References

Abdelgawad, W. (2012). The Bt Brinjal case: The first legal action against Monsanto and its Indian collaborators for biopiracy. *Biotechnology Law Report, 31*(2), 136–139.

African Union. (2018). *Continental strategy for Geographical Indications in Africa, 2018–2023*. Addis Ababa, Ethiopia: African Union Department of Rural Economy and Agriculture.

Agarwal, S., & Barone, M. (2005). *Emerging issues for Geographical indication branding strategies*. MATRIC Research Paper 05-MRP 9, January, Iowa State University.

Andersen, R. (2016). Historical context: Evolving international cooperation on crop genetic resources. In M. Halewood (Ed.), *Challenges in taxonomy and law. Farmers' crop varieties and Farmers' rights* (pp. 99–128). London: Taylor & Francis.

Arjun, K. M. (2013). Indian agriculture- status, importance and role in Indian economy. *International Journal of Agriculture and Food Science Technology, 4*(4), 343–346.

Barjolle, D. (2016). Geographical Indications and protected designations of origin: Intellectual property tools for rural development objectives. In D. S. Gangjee (Ed.), *Research handbook on intellectual property and Geographical Indications* (pp. 440–462). Cheltenham UK: Edward Elgar.

Becker, T. (2008). European food quality policy: the importance of geographical indications, organic certification and food quality insurance schemes in European countries. *The Estey Centre Journal of International Law and Trade Policy, 10*(1), 1–20.

Becker, T. (2009). European food quality policy: the importance of geographical indications, organic certification and food quality insurance schemes in European countries. *The Estey Centre Journal of International Law and Trade Policy, 10*(1), 111–130.

Belletti, G., et al. (2015). Linking protection of geographical indications to the environment: Evidence from the European Union olive-oil sector. *Land Use Policy, 48*, 94–106.

Bhalla, G. S., & Singh, G. (2001). *Indian agriculture: Four decades of development*. New Delhi, India: Sage.

Bhutani, S., & Kohli, K. (2012). Ten years of the biological diversity act. *Economic and Political Weekly, 47*(39), 15–18.

Blakeney, M. (2001). Intellectual property aspects of traditional agricultural knowledge. In M. Blakeney & P. Drahos (Eds.), *IP in biodiversity and agriculture: Regulating the biosphere* (pp. 29–52). London: Sweet & Maxwell.

Blakeney, M. (2002). Intellectual property aspects of traditional agricultural knowledge. In R. E. Evenson, V. Santaniello, & D. Zilberman (Eds.), *Economic and social issues in agricultural biotechnology* (pp. 43–60). Oxford, UK: CABI Publishing.

Blakeney, M. (2009). Protection of traditional knowledge by Geographical indications. *International Journal of Intellectual Property Management, 3*, 357–374.

Blakeney, M. (2016a). DNA patenting. In H. B. Singh, A. Jha, & C. Keswani (Eds.), *Intellectual property issues in biotechnology*. Wallingford, UK: CAB International.

Blakeney, M. (2016b). The negotiations in WIPO for international conventions on traditional knowledge and traditional cultural expressions. In J. C. Lai & A. M. Dominicé (Eds.), *Intellectual property and access to Im/material goods* (pp. 227–256). Cheltenham, UK: Edward Elgar.

Blakeney, M. (2017). Geographical Indications and environmental protection. *Frontiers of Law in China, 12*(2), 162–173.

Blakeney, M. (2019). *The protection of Geographical Indications. Law and practice* (2nd ed.). Cheltenham, UK: Edward Elgar.

Blakeney, M. Krishnankutty, J., Raju, R. K., & Siddique, K. H. M. (2019). Agricultural innovation and the protection of traditional rice varieties: Kerala a case study. *Frontiers in Sustainable Food Systems* (in print).

Blakeney, M., et al. (2012). *Extending the protection of Geographical Indications. Case studies in the protection of agricultural products in Africa*. London: Earthscan.

Brush, S. (1996). Whose knowledge, whose genes, whose rights? In S. B. Brush & D. Stabinsky (Eds.), *Valuing indigenous knowledge: Indigenous peoples and intellectual property rights* (pp. 1–24). Washington, DC: Island Press.

Chakravarti, A. K. (1973). The green revolution in India. *Annals of the Association of American Geographers, 63*(3), 319–330.

Chaudhary, R. C. (2019). Geographical Indications in Indian Agriculture. In C. Roy (Ed.), *The Role of Intellectual Property Rights in Agriculture and Allied Sciences* (pp. 241–271). Oakville, Canada: Apple Academic Press.

Correa, C. M. (2015). *Plant variety protection in developing countries. A tool for designing a sui generis plant protection system*. An alternative to UPOV 1991, Bonn, APREBES.

Cottier, T., & Panizzon, M. (2004). Legal perspectives on TK: the case for intellectual property protection. *Journal of International Economic Law, 7*(2), 371–399.

Cullet, P., et al. (2006). Intellectual property rights, plant genetic resources and traditional knowledge. In S. Biber-Klemm & T. Cottier (Eds.), *Rights to plant genetic resources and TK: Basic issues and perspectives* (pp. 112–154). London: CAB International.

Dagne, T. W. (2010). Harnessing the development potential of geographical indications for traditional knowledge-based agricultural products. *The Journal of World Intellectual Property, 5*(6), 441–458.

Das, A., et al. (2019). Farmers' rights and privileges: Implication to farming community. In C. Roy (Ed.), *The role of intellectual property rights in agriculture and allied sciences* (pp. 193–218). Oakville, Canada: Apple Academic Press.

Dattawadkar, N., & Mohan, H. (2012). *The status of geographical indications in India. A short review, India*. Jaipur, India: Intellectual Property Division, Stellarix Consultancy Services Pvt. Ltd..

Diallo, A. B. (2017). *Evaluation of the economic impact of geographical indications: Three case studies, Economics and Finance*. Clermont-Ferrand, France: Université Clermont Auvergne.

Downes, D. R. (2000). How intellectual property could be a tool to protect TK. *Columbia Journal of Environmental Law, 25*(2), 253–282.

Downes, D. R., & Laird, S. A. (1999). *Innovative mechanisms for sharing benefits of biodiversity and related knowledge: Case studies on Geographical Indications and trademarks*. Geneva, Switzerland: UNCTAD Biotrade Initiative.

FAO. (1994). Revision of the International Undertaking. Issues for consideration in stage II: access to plant genetic resources and Farmers' Rights, FAO Doc., CPGR-Ex1/94/5.

Gadgil, M. (2003). India's biological diversity act 2002: An act for the new millenium. *Journal of Biosciences, 28*(2), 145–147.

Gadwal, V. R. (2003). The Indian seed industry: Its history, current status and future. *Current Science, 84*(3), 399–406.

Gangjee, D. S. (2016). Genericide: the death of a Geographical indication? In D. S. Gangjee (Ed.), *Research handbook on intellectual property and geographical indications* (pp. 508–548). Cheltenham, UK: Edward Elgar.

Gerz, A., & Dupont, F. (2006). Comté cheese in France: Impact of a Geographical indication on rural development. In P. van de Kop, D. Sautier, & A. Gerz (Eds.), *Origin-based products: Lessons for pro-poor market development* (pp. 75–87). Amsterdam: KIT Publishers.

Giraud, G., & Amblard, C. (2003). What does traceability mean for beef meat consumer? *Sciences des Aliments, 23*(1), 40–46.

GIs Registry, Intellectual Property India. (2011). Certificate no. 158, Journal no 40 available at http://ipindiaservices.gov.in/GirPublic/Application/Details/199. Accessed 6 Dec 2019.

Glowka, L. (1998). *A guide to designing legal frameworks to determine access to genetic resources*. Gland, Switzerland: IUCN.

Helfer, L. (2007). The demise and rebirth of plant variety protection: A comment on technological change and the design of plant variety protection regimes'. *Chicago Kent Law Review, 82*, 1619–1627.

Hobbs, J. E. (2003). Information, incentives and institutions in the Agri-food sector. *Canadian Journal of Agricultural Economics, 51*(3), 413–429.

Hobbs, J. E., & Kerr, W. A. (2006). Consumer information, labelling and international trade in Agri-food products. *Food Policy, 31*(1), 78–89.

Hobbs, J. E., et al. (2005). Traceability in the Canadian red meat sector: Do consumers care? *Canadian Journal of Agricultural Economics, 53*(1), 47–65.

ISAAA. (2019). *Pocket K No. 35: Bt Brinjal in India*, available at https://www.isaaa.org/resources/publications/pocketk/35/default.asp. Accessed 6 Dec 2019.

Jagdish, J., Makanur, B., & Eraya. (2006, November 22–24). *Sustainable value of Navara Rice in Medical and Aesthetic Aspects, Agricultural Communication and Sustainable Rural Development: From Information to Knowledge to Wisdom-Envisioning a Food Sovereign World in the Third Millennium*, International seminar, Pantnagar, Uttarakhand.

Janis, M., & Smith, S. (2007). Technological change and the Design of Plant Variety Protection Regimes. *Chicago-Kent Law Review, 82*, 1557–1615.

Jebaraj, P. (2011, August 10). NBA for action against BT brinjal biopiracy, *The Hindu*, available at https://www.thehindu.com/news/national/nba-to-take-action-against-bt-brinjal-biopiracy/article2340768.ece

Kandavel, S. (2019, August 14). 'Panchamirtham' of Palani temple gets GI tag, *The Hindu.*, available at https://www.thehindu.com/news/national/tamil-nadu/panchamirtham-of-palani-temple-gets-gi-tag/article29089343.ece. Accessed 6 Dece 2019.

KAU (Kerala Agricultural University). (2018). *Wayanadan Nellinangal. Directory of farmers conserving traditional rice varieties*. Kerala, India: Government of Kerala, Department of Agricultural Development and Farmers Welfare, KAU.

Kochupillai, M. (2011). India's plant variety protection law: Historical and implementation perspectives. *Journal of Intellectual Property Rights, 16*(2), 1–21.

Kumar, S. (2011). Bt brinjal in India a long way to go. *GM Crops, 2*(2), 92–98.

Kumar, S., & Srivastava, S. (2017). The legal status of geographical indications in India. *Bioved, 28*(1), 43–56.

Kumar, S. and Srivastava, S. (2019) in Roy, C. (Ed)., *The role of intellectual property rights in agriculture and allied sciences* (pp. 66–98). Oakville Apple: Academic.

Lees, M. (2003). *Food authenticity and traceability*. Cambridge, UK: Woodhead Publishing.

Loureiro, M. L., & Umberger, W. J. (2007). A choice experiment model for beef: What US consumer responses tell us about relative preferences for food safety, country-of-origin labeling and traceability. *Food Policy, 32*(4), 496–514.

Louwaars, N. P., et al. (2005). *Impacts of strengthened intellectual property rights regimes on the plant breeding industry in developing countries: A synthesis of five case studies*. Wageningen, The Netherlands: Centre for Genetic Resources.

Lushington, K. (2012). The registration of plant varieties by farmers in India: A status report review of agrarian studies. *Review of Agrarian Studies, 2*(1), 112–128.

Maskus, K. E. (2000). *Intellectual property rights in the global economy*. Washington, DC: Institute for International Economics.

McCalman, P. (2001). Reaping what you sow: An empirical analysis of patent harmonization. *Journal of International Economics, 55*, 161–186.

McNeely, J. A. (2001). Biodiversity and agricultural development: The crucial institutional issues. In D. R. Lee & C. B. Barrett (Eds.), *Tradeoffs or synergies? Agricultural intensification, economic development and the environment* (pp. 399–417). Wallingford, UK: CABI.

Mengistie, G., & Blakeney, M. (2016). Geographical Indications in Africa. Opportunities, Experiences and Challenges'. *European Intellectual Property Review, 38*(5), 290–306.

Ministry of Agriculture. (1988). *New Seed Policy* (Letter No.11–71/88-SD-1 dated September 16, 1988) available at http://plantquarantineindia.nic.in/PQISPub/html/seepol.htm. Accessed 4 Dec 2019.

Ministry of Agriculture (India). (2002). *New Policy on Seed Development*, Letter No.11–71/88-SD-1 dated September 16, 1988, available at http://plantquarantineindia.nic.in/pqispub/html/seepol.htm. Accessed 4 Dec 2019.

Moschini, G., Menapace, L., & Pick, D. (2008). Geographical indications and the provision of quality in agricultural markets. *American Journal of Agricultural Economics, 90*(3), 794–812.

Murdoch, J., Marsden, T., & Banks, J. (2000). Quality, nature, and Embeddedness: Some theoretical considerations in the context of the food sector. *Economic Geography, 76*(2), 107–125.

Mutersbaugh, T., et al. (2005). Editorial. Certifying rural spaces: Quality certified products and rural governance. *Journal of Rural Studies, 21*(4), 381–388.

Nanda, N. (2013). *The protection of geographical indications in India: Issues and challenges*. New Delhi, India: TERI Briefing Paper.

O'Connor & Co. (2005). *Geographical indications and the challenges for ACP countries*. Brussels, Belgium: Agritrade.

Pai, Y., & Singla, T. (2016/2017). 'vanity GIs': India's legislation on geographical indications and the missing regulatory framework. In I. Calboli & N. Wee Loon (Eds.), *Geographical Indications at the crossroads of trade, Development and Culture* (pp. 333–358). Cheltenham, UK: Edward Elgar.

Pal, S., Tripp, R., & Louwaars, N. P. (2007). Intellectual property rights in plant breeding and biotechnology: Assessing impact on the Indian seed industry. *Economic and Political Weekly, 42*(3), 231–240.

Panizzon, M., & Cottier, T. (2005). Traditional knowledge and geographical indications: Foundations, interests and negotiating positions. In E. Petersmann (Ed.), *Developing countries in the Doha round: WTO decision-making procedures and WTO negotiations on trade in agricultural goods and services*. Florence, Italy: Robert Schuman Centre for Advanced Studies, European University Institute.

Parayil, G. (1992). The green revolution in India: A case study of technological change. *Technology and Culture, 33*(4), 737–756.

PPVFRA (Protection of Plant Varieties and Farmers' Rights Authority). (2018). *Compendium of registered varieties under PPV&FR Act, 2001*. New Delhi, India: Department of Agriculture and Co-operation, Ministry of Agriculture, Government of India.

Rana & Co. (2018). *India: Madhya Pradesh loses GI Tag, fails to join Basmati League*, available at https://www.lexology.com/library/detail.aspx?g=01a99c4e-688c-4370-9129-96e341af0df1

Ranjan, P. (2009). Recent developments in India's plant variety protection, seed regulation and linkages with UPOV's proposed membership. *The Journal of World Intellectual Property, 12*(3), 219–243.

Renting, H., Marsden, T. K., & Banks, J. (2003). Understanding alternative food networks: Exploring the role of short food supply chains in rural development. *Environment and Planning A, 35*(3), 393–411.

Réviron, S., Thevenod-Mottet, E., & El Benni, N. (2009). *Geographical indications: creation and distribution of economic value in developing countries*, NCCR Working Paper no 14.

Sagar, R. (2005). Intellectual property, benefit-sharing and traditional knowledge: How effective is the Indian Biological Diversity Act, 2002? *The Journal of World Intellectual Property, 8*(3), 383–404.

Sanderson, J. (2007). Back to the future: Possible mechanisms for the management of plant varieties in Australia. *University of New South Wales Law Journal, 30*(3), 686–702.
Sanderson, J., & Adams, K. (2008). Are plant Breeder's rights outdated? A descriptive and empirical assessment of plant Breeder's rights in Australia, 1987- Sanderson 2007. *Melbourne University Law Review, 32*(3), 980–1006.
Seema, P. S. (2012). *Protection of farmers' rights in India challenges for Law in the context of plant breeders' rights*, PhD thesis, Faculty of Law, Cochin University of Science and Technology.
Sen, A. (2019). IPR in agriculture and their prospects. In C. Roy (Ed.), *The role of intellectual property rights in agriculture and allied sciences* (pp. 1–64). Oakville, Canada: Apple Academic Press.
Shaji, K. A. (2018, May 29). Navara lands in IP rights row, *The Hindu*., available at https://www.thehindu.com/news/national/kerala/navara-lands-in-ip-rights-row/article24017583.ece. Accessed 3 Dec 2019.
Singh, H., & Aggarwal, R. (2013). Marketing of geographical indications in India: An analysis. *European Intellectual Property Review, 35*(11), 667–673.
Sood, J. (2015a, July 4). *Biopiracy case turns intense*, available at https://www.downtoearth.org.in/news/biopiracy-case-turns-intense-40323
Sood, J. (2015b, August 17). *Bt brinjal biopiracy case: Court vacates stay*, available at https://www.downtoearth.org.in/news/bt-brinjal-biopiracy-case-court-vacates-stay%2D%2D-42447. Accessed 6 Dec 2019.
Sreeja, V. N. (2013). Indian High Court reinstates criminal proceedings against Monsanto and its partners in India's first case of bio-piracy. *International Business Times*, 18 October, available at https://www.ibtimes.com/indian-high-court-reinstates-criminal-proceedings-against-monsanto-its-partners-1431534. Accessed 6 Dec 2012.
Srinivasan, C. S. (2004). Plant variety protection in developing countries. *Journal of New Seeds, 6*(1), 67–89.
Stanziani, A. (2004). Wine reputation and quality controls: The origin of the AOCs in 19th century France. *European Journal of Law and Economics, 18*(2), 149–167.
Stasi, A., et al. (2011). Italian wine demand and differentiation effect of geographical indications. *International Journal of Wine Business Research, 23*(1), 49–61.
Teuber, R. (2011). Consumers' and producers' expectations towards geographical indications; empirical evidence for a German case study. *British Food Journal, 113*(7), 900–918.
The Hindu, Kerala Tirur betel leaf, Tamil Nadu's Palani Panchamirtham get GI Tag, August 16, available at https://www.thehindubusinessline.com/news/kerala-tirur-betel-leaf-tamil-nadus-palani-panchamirtham-get-gi-tag/article29111502.ece. Accessed 6 Dec 2019.
Todhunter, C. (2019). Illegal Bt Brinjal Growing in India: A Call to Initiate Criminal Proceedings against Regulators and Corporations, *Asia-Pacific Research*, 15 May, available at https://www.asia-pacificresearch.com/illegal-bt-brinjal-growing-india-call-initiate-criminal-proceedings-against-regulators-corporations/5628637. Accessed 6 Dec 2019.
UPOV.UPOV Report on the Impact of Plant Variety Protection. Geneva.UPOV. 2005.
Van Der Ploeg, J. D., Renting, H., & Minderhoud-Jones, M. (2000). The socio-economic impact of rural development: Realities and potentials. *Sociologia Ruralis, 40*(4), 391–547.
Veena, P. G., & Rajasekharan, P. E. (2019). Access and benefit sharing in horticultural genetic resources. In P. E. Rajasekharan & V. R. Rao (Eds.), *Conservation and utilization of horticultural genetic resources* (pp. 637–662). Singapore, Singapore: Springer.
Verbeke, W., & Viaene, J. (1999). Consumer attitude to beef quality labeling and associations with beef quality labels. *Journal of International Food and Agribusiness Marketing, 10*(3), 45–65.
Vinayan, S. (2017). Geographical indications in India: Issues and challenges—An overview. *The Journal of World Intellectual Property, 20*(3 and 4), 119–132.
Winfree, J. A., & McCluskey, J. (2005). Collective reputation and quality. *Journal of Agricultural Economics, 87*(1), 206–213.
Wood, D., & Lenne, J. M. (1997). The conservation of agrobiodiversity on-farm: Questioning the emerging paradigm. *Biodiversity Conservation, 6*(1), 109–129.

World Bank. (2002). *World development report 2002*. Oxford, UK: Oxford University Press.
Wright, B. D. (1998). Intellectual property and farmers' rights. In R. Evenson, D. Gollin, & V. Santaniello (Eds.), *Agricultural values of plant genetic resources* (pp. 219–232). Wallingford, UK: FAO/CEIS/CABI.
WTO. (2019). *WTO members and observers*, available at https://www.wto.org/english/thewto_e/whatis_e/tif_e/org6_e.htm. Accessed 20 Dec 2019.
Wu, Y., & Zhang, Y. (2013). Analytical chemistry, toxicology, epidemiology and health impact assessment of melamine in infant formula: Recent progress and developments. *Food and Chemical Toxicology, 56*, 325–335.
Zhao, X., Finlay, D., & Kneafsey, M. (2014). The effectiveness of contemporary Geographical Indications (GIs) schemes in enhancing the quality of Chinese agrifoods – Experiences from the field. *Journal of Rural Studies, 36*, 77–86.

# Chapter 10
# Traditional Rice Cultivation in Kerala

**Michael Blakeney, Jayasree Krishnankutty, Rajesh K. Raju, and Kadambot H. M. Siddique**

**Abstract** This chapter looks at the cultivation of traditional rice varieties in Kerala, South West India. It seeks to test the effectiveness of the Protection of Plant Varieties and Farmers' Rights Act, 2001 and the Geographical Indications of Goods (Registration and Protection) Act 1999 in promoting agricultural innovation by reporting on a survey of 300 farmers of traditional rice varieties in Kerala. The study revealed that farmers were either unaware of the legislation, or unaware of its functions. They have not been much involved in the registration of farmers' varieties and have not made any benefit-sharing claims in relation to the varieties which have been registered. They have tended to confuse the registration of geographical indications with the registration of farmers' varieties. This suggests, as a first step, the necessity for awareness raising about the purposes of both pieces of legislation with Indian farmers.

**Keywords** Agricultural innovation · Traditional rice cultivation · Farmers' varieties · Kerala · Geographical indications · Intellectual property

This chapter builds upon M. Blakeney, J. Krishnankutty, R. K. Raju, and K. H. M. Siddique (2020) Agricultural Innovation and the Protection of Traditional Rice Varieties: Kerala a Case Study, *Frontiers in Sustainable Food Systems*, 3: article 116: 1–11).

M. Blakeney (✉)
School of Law, The University of Western Australia, Crawley, WA, Australia
e-mail: michael.blakeney@uwa.edu.au

J. Krishnankutty
Professor and Head, Communication Centre, Kerala Agricultural University, Thrissur, Kerala, India

R. K. Raju
Department of Agriculture Extension, College of Horticulture, Kerala Agricultural University, Thrissur, Kerala, India

K. H. M. Siddique
Hackett Professor of Agriculture Chair and Director of The UWA Institute of Agriculture, University of Western Australia, Crawley, WA, Australia



M. Blakeney, K. H. M. Siddique (eds.), *Local Knowledge, Intellectual Property and Agricultural Innovation*, https://doi.org/10.1007/978-981-15-4611-2_10

## 10.1 Rice Cultivation in India and Kerala

The evolution of domesticated Oryza sativa or Asian rice has been dated back approximately 9000 years (Oka, 1988). The wild ancestor *Oryza rufipogon* is identified as the source of the two major variety groups: *indica* and *japonica* (Londo et al., 2006). *Japonica* varieties were initially domesticated in southern China, around the Yangtze Valley (Purugganan, 2010). The cultivation of ancestral *indica* rice has been dated back to Neolithic times in the Ganges plains when it hybridized with domesticated *japonica* which had arrived from China (Khush, 1997). *Indica* consumption is estimated to have commenced by 8400 BP, with *indica* becoming a staple food by 5000 BP (Fuller, 2011; Gross & Zhao, 2014).

An astonishing diversity of rice varieties has been identified. Richharia and Govindasamy (1990) estimated that there were around 200,000 landraces of rice in India. Scented varieties were involved in religious rituals dating back 3000 years (Ahuja et al., 2008) and the Ayurvedic texts (Jose et al., 2018; Sathiya, 2013). These traditional varieties developed in a diverse range of landscapes across the subcontinent and with a considerable range of soil types, rainfall and micro-climates (Agnihotri & Palni, 2007; Jayashankar et al., 2001; Lakshmikutty, 2018). During the Green Revolution of the 1960s, high yielding varieties (HYV) of rice were developed which increased yield, reduced the cropping period and increased cropping intensity to allow the cultivation of 2–3 crops per year, but which required the use chemical fertilizers, pesticides, tractors, mechanical threshers and controlled water supply to crops (Janaiah & Debdutt, 2017). One of the major ecological consequences of the Green Revolution was the significant depletion in the number of traditional rice varieties as the HYV had a very narrow and unstable genetic base compared with traditional varieties (Ashraf & Lokanadan, 2017; Nelson, Ravichandran, & Antony, 2019; Parayil, 1992; Roy et al., 2019). Traditional varieties have gradually disappeared as farmers abandoned them in favour of monohybrid crops (Das & Das, 2014; Sathiya, 2013). In recent years and in the face of climate change it has been realised that traditional rice varieties represent a valuable gene pool for traits which may underpin the capacity of modern varieties of rice to adapt to climate change (Thrupp, 2000; Zhu et al., 2003).

This chapter examines the cultivation of traditional rice varieties in Kerala and the extent to which the laws described in the previous chapter support the conservation and utlilization of those varieties.

Kerala lies in the south-western corner of the Indian peninsula, in the southern part of the Western Ghats adjoining Tamil Nadu and Karnataka in the east and north-east, and bounded by the Arabian Sea in the west. Kerala is topographically and ecologically diverse, consisting of a mix of coastland, wetlands, and plains to the west, and the foothills of the Western Ghats to the east. The ecological conditions in the state have resulted in a considerable diversity of germplasm in both wild and cultivated rice.

Rice cultivation in Kerala dates back to 5000 BP. (Manilal, 1990). In addition to rice, the most common crops grown in Kerala are: tea, coffee spices, such as black pepper and cardamom, as well as areca nut and fruits such as bananas, coconut jackfruit and mango. Crop choice has depended primarily on topographical and climatic factors, but also on soil type, availability of irrigation, crop profitability, as well as upon agricultural policy (Guillerme et al., 2011; Kannan & Pushpangadan, 1990; Kumar, 2005; Narayanan, 2006). Archaeological evidence suggests that Kerala participated in global agricultural markets for at least 2000 years, trading spices first with the Romans, and later with Portuguese, Dutch, and British merchants (Jeffrey, 2001).

Rice is the staple food crop of Kerala, but Kerala has always experienced rice shortages. During 1960–1961, the peak period for rice production, the shortage of rice was 40.12% of the total demand and this increased to 83.45% in 2009–2010 (Karunakaran, 2014). Since the 1980s rice cultivation in Kerala has been in steady decline, from 8,500,000 hectares in 1980–1981 to 1,980,000 hectares in 2017 (GoK, 2017). The traditional rice growing areas in Kerala, Palakkad and Alappuzha, have, experienced 49.93% and 56.97% declines in the areas cultivated for rice between 1960–1961 and 2009–2010 (Karunakaran, 2014). The Palakkad district, which is considered the rice bowl of Kerala, has lost 1,03,980 hectares of paddy fields in the last four decades (Athira & Kumar, 2016). A number of factors have been identified as contributing to this decline, including: competition from other crops, such as rubber and coconut the difficulties involved in rice cultivation, such as biotic stress caused by diseases and weeds, low levels of productivity, uneven rainfall, land degradation, ground water depletion, chemical pollution and labour shortages (Athira & Kumar, 2016; Fox, 2017; Kannan, 1998; Kannan & Pushpangadan, 1990; Kumar, 2005; Mani, 2009). For example, Jose and Padmanabhan (2016) in a study of land use in the Wayanad district of Kerala, report that the subsistence wet-paddy integrated agro-ecosystem has given way to more lucrative cash crops, such as banana and ginger. Finally, the extensive conversion of paddy lands for residential purposes has also had a significant impact on the land available for rice farming in Kerala (Kumari, 2007).

A more recent factor contributing to the decline in global rice production is climate change. Rainfed cultivation is estimated to account for about 25% of global rice production, which makes it particularly vulnerable to fluctuations in rainfall, as well as heat stress from high temperatures (Soura et al., 2013; Singh, 2017).

With the expected demand for rice to increase in the coming years, food security will be imperilled, unless this situation can be improved. Farmers will have to increase yields by adopting high yielding varieties, or utilise those traditional varieties which are suitable for marginal lands.

## 10.2 Cultivation of Traditional Rice Varieties in Kerala

Latha et al. (2013) collected 623 different named rice landraces of Kerala, with 532 accessions characterized and conserved in the seed bank of National Gene Bank, National Bureau of Plant Genetic Resources, New Delhi. A survey conducted by the

Kerala Agricultural University (KAU, 2018). during November 2018 among 873 traditional rice farmers of the Wayanad district, identified 105 traditional varieties of rice. The Kerala Biodiversity Board, reported that out of the nearly 160 rice varieties of Wayanad, 55 traditional varieties are now extinct (PTI, 2012).

A study of traditional rice cultivation in the Palakkad, Malappuram and Wayanad districts, representing 47% of the total rice paddy area in Kerala (Krishnakutty et al., 2019) found that farmers were disinclined to cultivate traditional varieties because of their lower productivity in comparison with HYV, although the cost of traditional rice cultivation in all three districts was much lower than the state average for cultivating modern varieties. This was because the traditional varieties because these varieties have evolved locally and thrived for generations, resulting in fewer pest and diseases and the ability to withstand climatic variations. Their straw yield, relevant for cattle fodder, is high, as most of these cultivars are medium-tall to tall. The palatability, higher straw yield and straw preference by cattle are reasons expressed by farmers from other parts of India too, for preferring traditional rice varieties (Basu, 2017).

Krishnakutty et al. (2019) found that farm incomes were highest in Wayanad attributed to the market demand for three traditional varieties; *Navara* (medicinal variety) and *Gandhakasala and Jeerakasala* (aromatic varieties), which grow best in Wayanad and attract a higher price than other varieties, although famers reported the highest productivity for the traditional variety *Valichoori*. In Palakkad and Malappuram the main traditional varieties cultivated were *Chitteni, Chettadi, Thavalakkannan* and *Chenkazhama*.

The study revealed that most farmers in Kerala sell their produce through the state-owned procurement and distribution agency, Supplyco, for a pre-determined price, including a standard price for traditional rice, as fixed by the agency, irrespective of the variety. In Palakkad, Supplyco enjoyed an 80% share of the marketed volume, with the balance being sold to millers and friends and relatives In Wayand Supplyco sold 52.46% of the marketed volume, but *Jeerakasala* and *Gandhakasal* were sold through other channels to attract a premium price. In Malappuram a number of five marketing channels were identified, with Supplyco having the highest share of marketed volume, followed by *Nalla Bhakshana Prasthanam*, an NGO working to provide 'safe to eat' products. Farmers cultivating varieties such as *Navara* and *Rakthasali* were doing contract farming with a local miller who provided the seeds and procured the produce at a higher price than the prevailing local market price. The study indicated that traditional rice varieties are more prone to breakage during milling and have low hulling percentages than modern varieties, which. Reduces their commercial value and deterred wholesalers from procuring traditional varieties. Another reason identified for the low marketing efficiency of traditional rice is that farmers sell it as raw rather than dehusked grains because of the lack of suitable milling facilities for traditional rice. As a consequence, traditional rice is sold to wholesalers is mixed with modern varieties and sold under generic brand names.

Krishnakutty et al. (2019) found that the growers of traditional rice varieties tended to be older growers with a traditional disposition. They seemed to be unaware that they grow specialty rice and that they cultivated traditional varieties only as a continuation of ancestral practices. This is supported by the observations of Radhika (2014), Shamna (2014) and Rose (2011).

**Table 10.1** Registered farmers' rice varieties cultivated in Kerala

| Variety | Registration number |
|---|---|
| *Mullankayama (Mullanchanna)* | 572/2012 |
| *Thonnuran Thondi* | 573/2012 |
| *Kurumottam* | 576/2012 |
| *Kunjootti Matta* | 580/2012 |
| *Marathondi* | 583/2012 |
| *Onavattan* | 584/2012 |
| *Chenthandi* | 585/2012 |
| *Koduveliyan* | 588/2012 |
| *Thuroodi* | 589/2012 |
| *Valichoori* | 591/2012 |
| *Chennellu* | 56/2013 |
| *Gandhakasala* | 57/2013 |
| *Chomala* | 58/2013 |
| *Jeerakasala* | 59/2013 |
| *Veliyan* | 60/2013 |
| *Thondi* | 61/2013 |
| *Kottathondi* | 20 of 2016 |
| *Kayama* | 21 of 2016 |
| *Mannuveliyan* | 22 of 2016 |
| *Adukkan* | 23 of 2016 |

## 10.3 Protection of Traditional Rice Varieties in Kerala

As was mentioned in the previous chapter, the Government of India has sought to promote agricultural innovation, as well as the conservation of agricultural biodiversity by the passage of the Protection of Plant Varieties and Farmers' Rights Act, 2001 (PPVFR Act) and the Geographical Indications of Goods (Registration and Protection) Act, 1999 (GIs Act). A number of traditional rice varieties have been registered under these Acts.

## 10.4 Registration of Farmers' Varieties in Kerala

*The Plant Variety Rights Journal of India*, which is published by the Protection of Plant Varieties & Farmers' Rights Authority (PPVFRA), records the registration of 20 Farmers' Varieties of rice cultivated in Kerala (Table 10.1).

All of these registrations have been filed with the PPVFRA by the Secretary of Seed Care.[1] Seed Care describes itself as "an Association of Indigenous & Traditional

[1] Kunjootti Matta, Reg/2012/580, Marathondi, Reg/2012/583, Chenthadi, Reg/2012/585, Koduveliyan, Reg/2012/588, Thuroodi Reg/2012/589, Kurumottan. Reg/2012/576, Thonnuran

Crop Conservers of Malabar".[2] The Malabar region is the area of southwest India, including the state of Kerala along lying between the Western Ghats and the Arabian Sea. Seed Care has been operating since 2012 with the objectives of conserving and promoting the cultivation of traditional crop varieties in the Malabar region, protecting "farmer rights on seeds and associated knowledge systems" and building farmer networks concerned with agrobiodiversity conservation.[3] No information is given on the composition of Seed Care or its office holders.

The address given for Seed Care is "C/o M. S. Swaminathan Research Foundation, Community Agrobiodiversity Centre, Puthoorvayal, Wayanad, Kerala". The M S Swaminathan Research Foundation (MSSRF), was established in 1988 in Chennai, by the geneticist Professor M.S. Swaminathan as a not-for-profit trust. "aiming "to accelerate use of modern science and technology for agricultural and rural development to improve lives and livelihoods of communities".[4]

SEED CARE has explained that the registration of the famers' varieties of rice listed above was for the purposes of securing their availability for farmers; to instigate some pride among the farmers' by getting scientific validity to the varieties nurtured by them; and third, to attract breeders to access the scientifically validated varieties and accrue benefits for the farmers.[5] In 2011–2012, SEED CARE conducted a baseline survey in Wayanad to identify the traditional varieties of rice cultivated and area of cultivation, with special focus on speciality rice varieties (SEED CARE, 2012, p.48). Based on the survey, seeds of 10 speciality rice varieties were obtained for purification and four locations selected within the context of the national seed village programme, in which selected villages cultivate seed to be provided to neighbouring villages (India, 2002). In its 2013–2014 Annual Report SEED CARE listed the 10 purified varieties (SEED CARE, 2014, p.36).[6] During 2013–2014, a total of 3.15 tonnes of seeds was distributed to interested farmers, extending the existing area of 44.8 ha under traditional rice cultivation to a total of 74.8 ha (SEED CARE, 2014, p.36). In its 2014–2015 Annual Report the SEED CARE mentions the generation of 853 kg of purified seeds of nine traditional varieties[7] and its distribution to 54 farmers (SEED CARE, 2015, pp.35–36).

In relation to the marketing of traditional varieties of seed, the 2013–2014 Annual Report refers to a market study on the Gandhakasala variety was conducted with the help of Passau University, to look at the current status of the cultivation of the

Thondi, Reg/2012/573, Onavattan, Reg/2012/584, Taothabi Reg/2012/304, Gandha Malati Reg/2012/433, Valichoori Reg/2012/, Mullankayama (Mullanchanna) Reg/2012/572.

[2] https://www.mssrfcabc.res.in/programs/strengthening-grassroots-institution/seed-care/, accessed 1 November 2018.

[3] Ibid.

[4] http://www.mssrf.org/content/history-1, accessed 1 November 2018.

[5] Communication with MSSRF/Chennai/26-11-2018.

[6] Adukkan, Thondi, Mullan kayama, Gandhakasala, Jeerakasala, Chomala, Veliyan, Chennellu, Chenthadi, Kalladiaryan.

[7] Kalladiaryan (191 kg), Jeerakasala (106 kg), Chennellu (63 kg), Adukkan (135 kg), Chomala (76 kg), Thondi (40 kg), Veliyan (30 kg), Gandhakasala (178 kg) and Mullankaima (34 kg).

variety and to estimate the potential of collective marketing (SEED CARE, 2014, p.36). The study noted that the variety was mostly traded on the informal market, due to the lack of common procurement and processing and a uniform price and recommended the establishment of a "Farmers' Society/Consortium or a Producers' Company" and by the formation of Self-Help Groups (SHGs)/ Joint Liability Groups (JLGs) under the umbrella of an NGO (SEED CARE, 2014, p.36). Finally, the 2013–2014 Annual Report refers to efforts made for marketing of selected varieties under the brand name of "SEEDCARE" (SEED CARE, 2014, p.36).

The 2014–2015 Annual Report describes SEED CARE as a brand name for the marketing of traditional varieties of rice and mentions that Chennellu (red rice with medicinal value) was secured a rate of Rs. 25/kg as against the normal rate of Rs. 15/kg and Gandhakasala (aromatic variety) obtained Rs. 100/kg against the normal rate of Rs. 80/kg. (SEED CARE, 2015, p.36).

The Community Agricultural Biodiversity Centre (CAbC) was established in 1997 in Wayanad as one of a number of the regional centres of the MSSRF, confining its activities to the Western Ghats regions in Kerala.[8] The Centre describes itself as having been "established to promote community conservation systems of rural and tribal people through research, extension and advocacy" working "in partnership with rural and farming communities for sustainable agricultural and rural development".[9] The 2014–2015 Annual Report of SEED CARE Mentions the activities of the CAbC in the promotion of the marketing of traditional rice varieties through a "farmer -trader interface" and reports that 120 farm households benefited from the increased procurement price of rice. (SEED CARE, 2015, p.36).

In 2016 the CAbC assisted with the formation and registration of Wayanad Agri Marketing Producer Company Limited (WAMPCo), a farmer producer company named with the objectives of marketing traditional varieties of rice, vegetables, coffee and pepper and providing technical support to increase the productivity and quality of traditional crop varieties (SEED CARE, 2017, p.32).

Also mentioned in the Annual Report is the activity of the CAbC in the compilation of traditional and organic practices followed in rice cultivation in Wayanad (SEED CARE, 2015, p.36).

Given the general vague awareness, of rice farmers of the role and functions of the PPVFRA, disclosed in the surveys described below, it would seem that the SEED CARE and its associated institutions have assumed the primary role of biodiversity conservation envisaged by the legislation. It reported "Genome Saviour Awards" in 2008 and 2010 made by the PPVFR Authority to the Kurichya and Kuruma communities of Wayanad for conserving 20 traditional rice varieties with a range of characteristics, including tolerance to drought and flood, medicinal properties and aroma (SEED CARE, 2012, p.52).

[8] https://www.mssrfcabc.res.in/about-the-centre/, accessed 15 March 2019.

[9] Ibid.

The PPVFRA in s.26, together with rule 40 of the PPVFR Rules provides for inviting claims of benefit sharing in relation to varieties develop from registered varieties. There is no data on any benefit-sharing to date.

It should be noted that in addition to its rice conservation activities in Wayanad, the SEED CARE conducts similar activities in Chennai, where it conserves 500 accessions of different rice varieties at its Community Gene Bank, which have been multiplied in association with the Regional Rice Research Station, Tamil Nadu Agricultural University (TNAU) and at Tirur, Odisha in its Biju Patnaik Medicinal Plants Garden and Conservation Centre in Jeypore, where it has supplied seed materials of 75 traditional rice landraces and 27 popular rice landraces to central and state government institutions (SEED CARE, 2016, p.19).

### *10.4.1 Rice Registered Under the GIs Act*

To date 12 GIs have been registered for rice in India of which six are from Kerala: "Navara Rice", "Pokhali Rice", "Palakkadan Matta Rice", "Wayanad Jeerakasala Rice", "Wayanad Gandhakasala Rice "and "Kaipad Rice". In the cases of Navara and Pokhali rice, the GI is indirect, as the geographical origin has to be inferred from the name. The other six registered GIs for rice also include a number in which the geographical origin has to be inferred from the name: "Kalanamak Rice" (of Uttar Pradesh) "Ajara Ghansal Rice" and "Ambemohar Rice" (of Maharastra) "Gobindobhog Rice" and "Tulapanji Rice" (of West Bengal). Basmati Rice", as registered as a GI for rice produced in the states of Punjab, Haryana, Delhi, Himachal Pradesh, Uttarakhand, and parts of western Uttar Pradesh and Jammu & Kashmir.[10] Madhya Pradesh, Rajasthan and Bihar were excluded from this registration, on March 15, 2018 as not being in the traditional Basmati rice growing area in the Indo-Gangetic Plain' (Rana & Co, 2018). In relation to the final rice GI: "Joha rice of Assam", the geography is explicit.

The reputation of a product, being associated with a geographic area is usually established by resort to historical writings. In the application for the registration of "Navara Rice", reference was made to mention of the therapeutic qualities of the rice in the *Susruta Samhita* (2500 BCE) an the *Ashtanga Hriaya* (500 BCE)[11] The Statement of Case for "Palakkadan Matta rice" traces it to the times of the Cheras and Cholas (first to fourth century BCE) when the Palghat District, where it is grown was part of Tamil Nadu and is referred to in the Tamil classic *Tirukkural* (dated variously from 300 BCE to seventh century CE).[12] In the applications made for "Wayanad Jeerakasala Rice" and "Wayanad Gandhakasala Rice" reference is made

[10] Reg. No 145.

[11] Application form, available at http://ipindiaservices.gov.in/GI_DOC/17/17%20-%20Form%20GI-1%20-%2025-11-2004.pdf, accessed, 7 November 2018.

[12] *Statement of case*, available at http://ipindiaservices.gov.in/GI_DOC/81/81%20-%20Statement%20of%20Case%20-%2029-01-2007.pdf, accessed 5 November 2018.

to mentions of the cultivation of these rices in the "old verbal recitations in Malayalam called 'Krishippatu' describing the agricultural practices followed in Malayalakkara during the seventeenth century.[13] In the application for "Kaipad rice" it is conceded that the name "Kaipad" was not explicitly referred to in the ancient 'Kayal literature,[14] but was mentioned by Francis Buchanan in *A Journey from Madras through the Countries of Mysore, Canara, and Malabar*, which he undertook in 1801–1802.[15] Finally, in the application for "Pokkhali rice" the applicants refer to extracts from the Cochin State Manual published by the Cochin State Government in 1911, which contains "a detailed description of Pokkhali cultivation mentioning characteristics of traditional Pokkhali cultivars and its peculiar agro-climatic and soil characteristics".[16]

### 10.4.2 Registrants of Geographical Indications for Rice from Kerala

The GIs Act establishes a system for the registration of GIs. Section 11(1) of the GIs Act provides GIs may be registered by "any association of persons or producers or any organization or authority established by or under any law for the time being in force representing the interest of the producers of the concerned goods…" Generally, these applicants are involved in ensuring that farmers cultivating the varieties embraced by the GI registrations, adhere to prescribe cultivation and processing standards. This has the effect of preserving the commercial reputation of the GI. The registrants of the GIs for traditional rice varieties from Kerala are listed in the table below (Table 10.2).

Two varieties of *Navara* (black glumed and golden yellow glumed Navara) were registered as the GI "Navara Rice" by the Navara Rice Farmers Society, Karukamanikalam, near Chittur (The Hindu, 2008).

The Pokkali Land Development Society and Kerala Agricultural University (KAU) were joint applicants for the GI "Pokkali Rice" (The Hindu, 2006a, 2006b).

KAU and the Wayanad Zilla Nellulpadaka Karshaka Samithi (a farmers' collective), were joint applicants for the GIs "Wayanad Jeerakasala Rice" and "Wayanad Gandhakasala Rice" (The Hindu, 2010).

Malabar Kaipad Farmers' Society (MKFS) of Ezhome obtained the registration of the GI "Kaipad Rice" (Nazeer, 2014). The society was formed for the promotion

---

[13] *Statement of case*, available at http://ipindiaservices.gov.in/GI_DOC/186/186%20-%20 Statement% 20of%20Case%20-%2023-09-2009.pdf, accessed 5 November 2018.

[14] Statement of case, http://ipindiaservices.gov.in/GI_DOC/242/242%20-%20GI%20-%20 Reply%20to%20 ER%20-%20Statement%20of%20Case%20-%2007-10-2013.pdf, accessed 5 November 2018.

[15] reprinted by Cambridge University Press, 2012.

[16] Statement of case available at http://ipindiaservices.gov.in/GI_DOC/81/81%20-%20 Statement%20of%20 Case%20-%2029-01-2007.pdf, accessed 5 November 2018.

**Table 10.2** Registered Geographical Indications of Traditional Rice from Kerala

| Cert. no | Geographical indication | Applicant | Date available |
|---|---|---|---|
| 40 | Navara Rice | Navara Rice Farmers Society<br>Navara Eco Farm, Karukamanikalam, Chittur College, P.O., Palakkad – 678 104, Palakkad , Kerala | 20/06/2007 Until 24/11/2024 |
| 41 | Palakkadan Matta Rice | Palakkadan Matta Farmers Producer Company Limited<br>Karukamanikalam,<br>Chittur College P.O., Palakkad – 678104 | 20/06/2007 Until 17/04/2025 |
| 81 | Pokkali Rice agricultural | (i) Kerala Agricultural University<br>P.O. Thrissur District, Kerala – 680 656 (ii) Pokkali Land Development Agency, N. Paravur, Ernakulam District, Kerala | 26/05/2008 Until 28/01/2027 |
| 137 | Wayanad Jeerakasala Rice | (i) Kerala Agricultural University and (ii) Jilla Sugandha Nellulpadaka Karshaka Samithi, Rural Agricultural Wholesale Market, Sulthan Bathery, Wayanad – 673 592, Kerala. | 31/05/2010 Until 21/09/2019 |
| 138 | Wayanad Gandhakasala Rice | (i) Kerala Agricultural University and Wayanad; (ii) Jilla Sugandha Nellulpadaka Karshaka Samithi | 31/05/2010 Until 21/09/2019 |
| 242 | Kaipad Ricel | (i) Malabar Kaipad Farmers' Society Ezhome Grama Panchayat, Ezhome P.O, Kannur – 670 334, Kerala,. (ii) Kerala Agricultural University. | 30/10/2013 |

Source 'Registered GIs' Geographical Indications Registry available at http://ipindiaservices.gov.in/GirPublic/Application/Details/81, accessed 5 November 2018

of 'kaipad' farming in Kannur, Kasaragod, and Kozhikode on the initiative of the College of Agriculture at Padannakkad in Kasaragod (Nazeer, 2014).

"Palakkadan Matta Rice" was registered by the Palakkad Matta Farmers Producer Company Ltd. (The Hindu, 2008). The Palakkaddan matta is described as bold red rice with a unique taste because of its special geographical area and peculiar weather of Eastern wind. Other rice varieties with matta properties cultivated in Palakkad can be added to this list after detailed examination (The Hindu, 2008).

The "Palakkadan Matta Rice" registration was obtained by the Palakkad Matta Farmers Producer Company Ltd., whose Chairman was Mr. P. Narayanan Unny, the President of the Navara Rice Farmers Society. The company comprised 10 of the 5000 producers of the varieties embraced by the registration (Marie-Vivien, 2015).

Soam (2005) mentions the symbiotic relationship between Pokkali rice and prawn production in the flooded paddy fields and that paradoxically, the greater profitability of prawns is causing farmers to abandon rice production. Soam (2005) also mentions the potential for Jeerakasala and Gandhakasala rice, scented varieties grown organically in the Wayanad District of the Kerala State, especially by the Kurichiyas tribe, to be marketed by GIs.

The applicant for the "Navara" GI was the Navara Rice Farmers Society, at Karukamanikalam, near Chittur. Its President, Mr. P. Narayanan Unny, was the proprietor of the Navara Eco Farm, at which purification of the Navara variety had been undertaken since 1994 (Priyadershini, 2018). Mr. Unny, had apparently sought to register the GI with three farmers from his farm, but this had been rejected by the GIs Registry which said that it was not prepare to accept a GIs registration from a single farm and the Navara Rice Farmers Society, was established to overcome this difficulty (Marie-Vivien, 2015). The Registry sought assurances that the interests of other Navara rice growers would be represented by the Society (Marie-Vivien, 2015, text at n.24). The applicant consulted with stakeholder farmers, the Kerala Agriculture University, rice millers and traders (Priyadershini, 2018) and the assistance of the National Bank for Agriculture and Rural Development (NABARD) was obtained for seed purification, multiplication and expansion of the area of cultivation (*The Hindu*, 2010).

The Pokkali Land Development Society and Kerala Agricultural University (KAU) were joint applicants for the GI "Pokkali Rice" (The Hindu, 2006a, b). They are also the inspection bodies named in the registration.[17]

KAU and the Wayanad Zilla Nellulpadaka Karshaka Samithi (a farmers' collective), were joint applicants for the GIs "Wayanad Jeerakasala Rice" and "Wayanad Gandhakasala Rice" (The Hindu, 2010).

The Malabar Kaipad Farmers' Society (MKFS) of Ezhome obtained the registration of the GI "Kaipad Rice" (Nazeer, 2014). The society was formed for the promotion of 'Kaipad' farming in Kannur, Kasaragod, and Kozhikode on the initiative of the College of Agriculture at Padannakkad in Kasaragod (Nazeer, 2014).

## 10.5 Relationship Between Rice Registered Under the PPVFRA and the GIs Act

The existence of two separate pieces of legislation applying to traditional rice varieties has resulted in a degree of confusion (see Blakeney, et al., 2020). The PPVFRA is concerned with the registration of farmers' varieties and the GIs Act is concerned with the designations under which varieties are marketed. Confusion may arise for because a number of different varieties of rice can be embraced by a single GI. For example, the registration of the GI "Navara rice" covers two varieties of Navara: black glumed and yellow glumed. The registration of the GI "Palakkadan Matta" include 10 varieties: Aryan, Aruvakkari, Chitteni, Chenkazhama, Chettadi, Thavalakanna, Eruppu, Poochamban, Vattan Jyothy and Kunjukunj (The Hindu, 2008). and the registration permits the addition to this list of more rice varieties with matta properties and cultivated in Palakkad can be added after examination (see discussion in Kochar, 2008, p.341). However, it should be noted that the GIs registration only concerns the right to use the registered designation in marketing and does not affect the right of farmers to cultivate the varieties which are included in a registered designation. The registration of "Palakkadan Matta Rice" as a GI was

[17] Statement of case available at http://ipindiaservices.gov.in/GI_DOC/81/81%20-%20Statement%20of%20Case%20-%2029-01-2007.pdf, accessed 5 November 2018.

apparently obtained in the face of opposition from the scientific community in the state and the Department of agriculture “refuted any link between geography and palakkad matta rice” (ICAR, 2007). Additionally, the millers who have major say in deciding the market price of rice claim that the registration was not going to increase the price of Palakkadan matta and claimed that the quality of the rice came from milling and had no link to the geography (ICAR, 2007). An academic study undertaken in Chittur taluk of Palakkad district, (Rose, 2011) concluded that the impact of GI registration “was marginal in terms of increase in annual agricultural income and possession of farm and household assets” but that “GI registration was partially successful in securing higher price (Rs. 14.01/kg), maintaining area of cultivation and increasing institutional participation among farmers” and that consumers preferred the GI rice. On the other hand, an UNCTAD study (Das, 2009) disclosed that revealed that origin guaranteed agricultural products could secure a price premium in India of between 10 and 15%.

In a study of the attitude of farmers to the registration of the GI for Pokkali rice, Anson & Pavithran, (2014) ‘suggest a generally indifferent or negative reaction of farmers, with a threat to the sustainability of the production of the rice over the next 10–20 years because the new generation was not willing to continue with for Pokkali rice production. They also suggest that rethinking of the GI Act is necessary as although the economic value of the GI products in the market is very high, supply chain management is not in the hands of the producers and thus the major profit goes to intermediaries.

An illustration of some of the confusion surrounding the legislation protecting farmers’ varieties is a report that on 1 June 2018 that the state government of Kerala and the Kerala Agriculture University had objected to a petition filed by a farmer from Palakkad for registering “Navara” under the PPVFR Act (Sushma, 2018). The basis of the objection was that as Navara had been cultivated for centuries it was “not ethical to patent it under a single farmer’s name” (Sushma, 2018). This report also illustrates some confusion about the effect of the registration of a farmers’ variety, which has nothing to do with patenting.

Interestingly, the applicant for registration of Navara as a farmers’ variety under the PPVFR Act. had already secured registration of Navara as a GI under the GIs Act. This registration did not confer exclusive marketing or cultivation rights upon the registrant, but merely protected the right of farmers in the geographical area associated with Navara cultivation, to use the designation in the marketing of their products.

It should be noted that the PPVFR Act provides in s.15(4) that a new variety shall not be registered under the Act “if the denomination given to such variety— (viii) is comprised of solely or partly of geographical name”. However, a proviso to s.15(4) states that “that the Registrar may register a variety, the denomination of which comprises solely or partly of a geographical name, if he considers that the use of such denomination in respect of such variety is an honest use under the circumstances of the case.”

In relation to rice cultivation in Kerala, the experience seems to suggest that the separate statutes which purport to encourage agricultural innovation appear to be somewhat contradictory in their operation and do not have the full support of beneficiary communities.

A recent study of the impact of GIs on the well-being of rice farmers in Kerala confirmed that a price premium could be secured for rice marketed under its GI (Radhika et al., 2018). It noted that the financial returns were greatest for Navara followed in order by Palakkadan Matta, Gandhakasala, Kaipad, Jeerakasala and Pokkali and that this was attributed to the energy involved in the marketing of the different rice types under their GI (Radhika et al., 2018).

In a study of the attitude of farmers to the registration of the GI for *Pokkali* rice, (Anson and Pavithran, 2014) suggests a generally indifferent or negative reaction of farmers, with a threat to the sustainability of the production of the rice over the next 10–20 years because the new generation was not willing to continue with for *Pokkali* rice production. They also suggest that rethinking of the GI Act is necessary as although the economic value of the GI products in the market is very high, supply chain management is not in the hands of the producers and thus the major profit goes to intermediaries.

Soam (2005) mentions the potential for *Jeerakasala* and *Gandhakasala* rice, scented varieties to be marketed by GIs. Soam (2005) mentions the symbiotic relationship between *Pokkali* rice and prawn production in the flooded paddy fields and that paradoxically, the greater profitability of prawns is causing farmers to abandon rice production.

## 10.6 Conclusions

Even though Kerala has the highest (Human Development Index (HDI) in India (0.72 in 2015) and literacy rate (93.91 in the 2011 census), the farmers surveyed in Blakeney et al. (2020) disclosed an imperfect and vague knowledge of the functions and details of the PPVFR and GIs Acts. For example, the protection of the varieties *Chitteni Chettadi* and *Thavalakkannan* were erroneously reported by the farmers surveyed as being protected by a geographical indication. Also unclear on the part of respondents was who benefitted from the two different kinds of protection and the effects of registration.

The registration record discloses that all of the farmers' varieties from Kerala registered under the PPVFR Act were obtained by the M S Swaminathan Research Foundation. As it mentioned, its primary motivations for these registrations were to preserve biodiversity and to promote sustainable agriculture (Swaminathan, 2018). There is no indication as to whether the registered farmers' varieties have contributed to the development of new rice varieties. There is also no evidence of any attempt by SEED CARE or farmers from Kerala to seek any benefit- sharing in relation to use of the registered farmers' varieties in the development of new varieties.

There is no indication as to the reasons for the selection by SEED CARE from these varieties of the 15 which they have registered. The SEED CARE 2014–2015

Annual Report refers to "SEED CARE" as a brand name for the marketing of traditional varieties of rice and *Chennellu* and *Gandhakasala* as priority varieties for this marketing (SEED CARE, 2015, 36). Its marketing of *Gandhakasala*, might bring it into conflict with KAU and Jilla Sugandha Nellulpadaka Karshaka Samithi, which have secured registration of the geographical indication "Wayanad Gandhakasala Rice". The marketing by SEED CARE of *Chennellu* might cause difficulties with the Palakkad Matta Farmers Producer Company Ltd., which has included the variety in its registered geographical indication: "Palakkadan Matta Rice".

A number of farmers surveyed expressed some skepticism about the usefulness of geographical indications in securing higher prices for Palakkadan Matta Rice (see also Ajayan, 2009) Mr. P. Narayanan Unny, the President of the Navara Rice Farmers Society, was quoted as saying that there was a market for this rice among the Keralite population in West Asia, Europe and the USA, but that "the GI status we earned after years of work has not added any flavour to the lives of farmers as we expected" (Ajayan, 2009).

As the cultivation of traditional rice varieties is dependent on the price received, the use of GIs. will help the realization of premium prices and attract more farmers to traditional rice cultivation (Radhika, Thomas, Kuruvila, & Raju, 2018) However, a number of the farmers surveyed identified high labour costs as outweighing the returns from price premiums for rice sold under geographical indications (see also Ajayan, 2009).

The protection and marketing of farmer varieties of rice is a matter of crucial importance in a state like Kerala, which is a representative of a modern agricultural state in an advanced developing country. Although the PPVFR Act and the GIs Act represent legislative initiatives of the Indian Government, designed and advanced for sustaining traditional agriculture, they appear to be unnoticed by the target beneficiaries. The different objectives of the two Acts are unclear, and they are not considered to be user friendly. Agricultural extension programmes with the assistance of legal could address this situation.

Finally, it should be acknowledged that in Kerala, as in many other developing countries, agriculture is a way of living and a continuity of tradition for the average farmer, who has traditionally survived on subsistence production (Kwa, 2001). Viewing agricultural production and marketing as entrepreneurial activities appears to be alien to most farmers. At the same time the increasing urbanization and diversification of occupations in Kerala is relegating traditional rice cultivation to a less significant position in the state economy. Given the importance of the genetic diversity of traditional rice varieties, particularly at a time of climate change, governments as a matter of policy might consider the introduction of financial and other incentives to encourage the on-farm conservation of these varieties (Prakash, et al., 2007).

## References


Agnihotri, R. K., & Palni, L. M. S. (2007). On-farm conservation of landraces of rice (Oryza sativa L.) through cultivation in the Kumaon Region of Indian Central Himalaya. *Journal of Mountain Science, 4*(4), 354–360.

Ahuja, U., et al. (2008). Scented Rices of India. *Asian Agri-History, 12*(4), 267–283.

Ajayan. (2009). *Farmers yet to benefit from GI tag*, available at https://www.livemint.com/Money/EfOjbzNUEKv6zVOjBZvSPP/Farmers-yet-to-benefit-from-GI-tag.html. Accessed 9 Nov 2018.

Anson, C. J., & Pavithran, K. B. (2014). Pokkali Rice production under geographical indication protection: The attitude of farmers. *Journal of Intellectual Property Rights, 19*, 49–53.

Ashraf, A. M., & Lokanadan, S. (2017). A review of Rice landraces in India and its inherent medicinal values -the nutritive food values for future. *International Journal of Current Microbiology and Applied Science, 6*(12), 348–354.

Athira, H., & Kumar, N. K. (2016). Scenario analysis of rice cultivation in Kerala. *Journal of Extension Education, 28*(4), 5760–5763.

Basu, M. (2017). Indigenous appeal: Farmers in West Bengal are going back to indigenous rice varieties, with a little encouragement, *DownToEarth*, 18 July, available at: https://www.downtoearth.org.in/news/agriculture/indigenous-appeal-58285. Accessed 16 Mar 2019.

Blakeney, M., Krishnankutty, J., Raju, R. K., & Siddique, K. H. M. (2020). Agricultural innovation and the protection of traditional rice varieties: Kerala a case study. *Frontiers in Sustainable Food Systems, 3*(article 116), 1–11.

Das, K. (2009). *Socio-Economic implications of protecting geographical indications in India, Centre for WTO Studies*. Available online at: http://wtocentre.iift.ac.in/papers/Gi_Paper_CWS_August%2009_Revised.pdf. Aaccessed 30 April 2020.

Das, T., & Das, A. K. (2014). Inventory of the traditional rice varieties in farming system of southern Assam: A case study. *Indian Journal of Traditional Knowledge, 13*(1), 157–163.

Fox, T. A. (2017). Agricultural land-use change in Kerala, India: Perspectives from above and below the canopy. *Agriculture, Ecosystems & Environment, 245*, 1–10.

Fuller, D. Q. (2011). Finding plant domestication in the Indian subcontinent. *Current Anthropology, 52*(S4), S347–S362.

GoK (Government of Kerala). (2017). *Farm Guide*. Farm Information Bureau, Thiruvananthapuram, GoK, cited in Athira & Kumar, 2016.

Gross, B. L., & Zhao, Z. (2014). Archaeological and genetic insights into the origins of domesticated rice. *Proceedings of the National Academy of Sciences of the United States of America, 111*(17), 6190–6197.

Guillerme, S., Kumar, B. M., Menon, A., et al. (2011). Impacts of public policies and farmer preferences on agroforestry practices in Kerala, India. *Environmental Management, 48*, 351–364.

ICAR. (2007). *Geographical indications mapping: Palakkad Matta Rice draft report of field experience training: 81 FOCARS (25th August to 7th September 2007)*, Hyderabad National Academy of Agricultural Research Management.

Janaiah, A., & Debdutt, B. (2017). The Rice seed system in India: Structure, performance, and challenges, intellectual property rights, innovation and Rice strategy for India. In S. Mohanty et al. (Eds.), *The future Rice strategy for India* (pp. 359–382). London: Elsevier Science & Technology.

Jayashankar, N., et al. (2001). *Indigenous rice varieties*. Chennai, India: Chennai Centre for Indian Knowledge Systems.

Jeffrey, R. (2001). *Politics, women and well-being: How Kerala became a model*. Oxford: Oxford University Press.

Jose, M., & Padmanabhan, M. (2016). Dynamics of agricultural land use change in Kerala: A policy and social-ecological perspective. *International Journal of Agricultural Sustainability, 14*(3), 307–324.

Jose, M., et al. (2018). The Prehistoric Indian Ayurvedic Rice Shashtika is an extant early domesticate with a distinct selection history. *Frontiers in Plant Science*, *9*: (Article 1203).

Kannan, K. P. (1998). Political economy of labour and development in Kerala. *Economic and Political Weekly, 33*(52), L61–L70.

Kannan, K. P., & Pushpangadan, K. (1990). *Dissecting agricultural stagnation in Kerala: An analysis across crops, seasons and regions*, CDS working paper, 238. Trivandrum, India: CDS.

Karunakaran, N. (2014). Paddy cultivation in Kerala – Trends, determinants and effects on food security. *Artha – Journal of Social Sciences, 13*(4), 21–35.

KAU. (2018). *Wayanadan Nellinangal. Directory of farmers conserving traditional rice varieties*, Govt of Kerala Dept of Ag. Dev. and Farmers Welfare – Kerala Agricultural University.

Khush, G. S. (1997). Origin, dispersal, cultivation and variation of rice. *Plant Molecular Biology, 35*, 25–34.

Kochar, S. (2008). Institutions and capacity building for the evolution of intellectual property rights regime in India: IV– identification and disclosure of IP products for their IPR protection in plants and animals. *International Journal of Intellectual Property Rights, 13*, 336–343.

Krishnakutty, J., et al. (2019). *Traditional Sustainability of Rice Cultivation in Kerala*, India-A Socio-economic Analysis, (submitted for publication).

Kumar, B. M. (2005). Land use in Kerala: Changing scenarios and shifting paradigms. *Journal of Tropical Agriculture, 42*(1-2), 1–12.

Kumari, S.L. (2007). *Status paper on rice. Rice knowledge management portal (RKMP)*. Directorate of Rice Research, Rajendranagar, Hyderabad, available at. http://www.rkmp.co.in/sites/default/files/ris/rice-state-wise/Status%20Paper%20on%20 Rice%20in%20 Kerala.pdf. Accessed 6 Dec 2019.

Kwa, A. (2001). *Agriculture in Developing Countries: Which Way Forward? South Centre, Trade-Related Agenda*. Development and Equity (T.R.A.D.E.) Occasional Papers, 4. https://focus-web.org/publications/2001/agriculture_which_way_forward.html. Accessed 26 Jan 2019.

Lakshmikutty, D. (2018). *Traditional Rice varieties of India*, available at https://www.esamskriti.com/e/Culture/Indian-Culture/Traditional-Rice-Varieties-of-India-1.aspx. Accessed 6 Dec 2019.

Latha, M., et al. (2013). Rice landraces of Kerala state of India: A documentation. *International Journal of Biodiversity and Conservation, 5*(4), 250–263.

Londo, J. P., et al. (2006). Phylogeography of Asian wild rice, Oryza rufipogon, reveals multiple independent domestications of cultivated rice, Oryza sativa. *Proceedings of the National Academy of Sciences of the United States of America, 103*, 9578–9583.

Mani, K. P. (2009). Cropping pattern in Kerala– Spatial inter-temporal analysis. In K. Rajan (Ed.), *Kerala economy: Trends during the post-reform period* (pp. 64–84). New Delhi, India: Serials Publications.

Manilal, K. S. (1990). Ethnobotany of the Rices of Malabar. In S. K. Jain (Ed.), *Contribution to Ethnobotany of India* (pp. 243–253). Jodhpur, India: Scientific Publishers.

Marie-Vivien, D. (2015). *The protection of geographical indications in India. A New Perspective on the French and European Experience*. New Delhi, India: Sage.

Narayanan, N. C. (2006). For and against grain land use politics of rice in Kerala, India. *International Journal of Rural Management, 2*(1), 123–144.

Nazeer, M. (2014). “Kaipad” rice included in GI registry. *The Hindu*, June 9, available at https://www.thehindu.com/todays-paper/tp-national/tp-kerala/kaipad-rice-included-in-gi-registry/article6096417.ece. Accessed 18 July 2018.

Nelson, A. R. L. E., Ravichandran, K., & Antony, U. (2019). The impact of the green revolution on indigenous crops of India, *J Ethnic Foods*, 6: Article no.8.

Oka, H. (1988). *Origin of cultivated rice*. Tokyo: Elsevier Science.

Parayil, G. (1992). The green revolution in India: A case study of technological change. *Technology and Culture, 33*(4), 737–756.

Prakash, T. N., et al. (2007). Farmers' willingness to conserve crop genetic resources under different policy incentives. The case of traditional rice varieties in the Western Ghats of South India. *Outlook on Agriculture, 36*(2), 137–143.
Priyadershini, S. (2018). *An old Kerala family farm is reviving the near-forgotten navara rice variety*. Available at: https://www.thehindu.com/life-and-style/food/an-old-kerala-family-farm-is-reviving-the-near-forgotten-navara-rice-variety/article24652765.ece. Accessed 9 Nov 2018.
PTI (Press Trust of India). (2012). *Traditional rice varieties vanishing in Kerala*, available at https://www.thehindubusinessline.com/economy/agri-business/Traditional-rice-varieties-vanishing-in-KeralaStudy/article 20476098.ece. Accessed 6 Dec 2019.
Purugganan, M. D. (2010). The evolution of rice: Molecular vignettes on its origins and spread. *Archaeological and Anthropological Sciences, 2*, 61–68.
Radhika, A. M. (2014). *Economic analysis of production and marketing of Kaipad paddy in Kannur District*. M.Sc (Ag) thesis, Kerala Agricultural University, Thrissur, 130p.
Radhika, A. M., Thomas, K. J., Kuruvila, A., & Raju, R. K. (2018). Assessing the impact of geographical indications on well-being of Rice farmers in Kerala. *International Journal of Intellectual Property Rights, 9*(2), 1–11.
Rana, S. S & Co. (2018). *India: Madhya Pradesh Loses GI Tag, Fails to Join Basmati League*. Available online at: https://www.lexology.com/library/detail.aspx?g=01a99c4e-688c-4370-9129-96e341af0df1. Accessed 30 April 2010.
Richharia, R. H., & Govindasamy, S. (1990). *Rices of India*. Karjat, India: India Academy of Development Science.
Rose, N. (2011). *Impact of Geographical Indication (GI) on Palakkadan Matta Rice farmers in Kerala – An economic analysis*. Thesis submitted to the University of Agricultural Sciences, Bangalore in partial fulfillment of the award of the Degree of Master of Science (Agriculture) in Agricultural Economics, Bangalore.
Roy, K., et al. (2019). Protecting non-Basmati Indigenous Aromatic Rice varieties of West Bengal, India under Geographical Indication: A critical consideration. In C. Roy (Ed.), *The role of intellectual property rights in agriculture and allied sciences* (pp. 273–295). Oakville, Canada: Apple Academic Press.
Sathiya, A. (2013). Are the Indian Rice landraces a heritage of biodiversity to reminisce their past or to reinvent for future? *Asian Agri-History, 17*(3), 221–232.
SEED CARE. (2012). *Twenty-Second annual report, 2011–2012*, Chennai, SEED CARE, available at http://59.160.153.188/library/sites/default/files/AR2011-2012.pdf. Accessed 5 Nov 2018.
SEED CARE. (2014). *Twenty-Fourth annual report, 2013–2014*, Chennai, SEED CARE, available at http://www.SeedCare.org/SeedCareoldsite/sites/default/files/SEEDCARE_Annual%20Report_2013-14.pdf. Accessed 5 Nov 2018.
SEED CARE. (2015). *Twenty-Fifth Annual Report, 2014–2015*, Chennai, SEED CARE, available at http://www.SeedCare.org/SeedCareoldsite/sites/default/files/Annual%20Report%202015.pdf. Accessed 5 Nov 2018.
SEED CARE. (2016). *Twenty-Sixth Annual Report, 2015–16*, available at http://www.SeedCare.org/SeedCareoldsite/sites/default/files/Annual%20Report%202016.pdf. Accessed 5 Nov 2018.
SEED CARE. (2017). *Twenty-Seventh annual report, 2016–17*, Chennai, SEED CARE, available at http://59.160.153.188/library/sites/default/files/Annual%20Report%202016-17.pdf. Accessed 5 Nov 2018.
Singh, K. (2017). Mapping regional risks from climate change for rainfed rice cultivation in India. *Agricultural Systems, 156*, 76–84.
Shamna, N. (2014). *A study on farmers perceptions on prospects and problems of Pokkali rice farming in the State of Kerala*. (M.Sc (Ag) thesis). Hyderabad, India: Professor Jayshankar Telengana State Agricultural University.
Soam, S. K. (2005). Analysis of prospective geographical indications of India. *The Journal of World Intellectual Property, 8*(5), 679–706.
Soura, N. K., et al. (2013). An assessment of regional vulnerability of rice to climate change in India. *Climatic Change, 118*(3-4), 683–699.

Sushma, M. (2018). *Navara rice controversy: Plant variety act doesn't protect farmers' interests, say experts, down to earth*. Available online at: https://www.downtoearth.org.in/news/navara-rice-controversy-plant-varietyact-does-not-protect-farmers-interests-say-experts-60728. Accessed 30 April 2020.

Swaminathan Foundation. (2018). *'On the Farmers' rice varieties of Wayanad, Kerala, correspondence with the Authors, 26 November (Chennai)*.

The Hindu. (2006a). *Initiative to get GI registration for Pokkali rice by this month*, available at https://www.thehindu.com/todays-paper/tp-business/Initiative-to-get-GI-registration-for-Pokkali-rice-by-this-month/article15734656.ece. Accessed 18 July 2018.

The Hindu. (2006b). *Initiative to get GI registration for Pokkali rice by this month*, available at https://www.thehindu.com/todays-paper/tp-business/Initiative-to-get-GI-registration-for-Pokkali-rice-by-this-month/article15734656.ece. Accessed 3 Dec 2019.

The Hindu. (2008, March 17). *Recognition for two rice varieties.*, available at https://www.thehindu.com/todays-paper/tp-national/tp-kerala/Recognition-for-two-rice-varieties/article15185764.ece. Accessed 3 Dec 2019.

The Hindu. (2010, November 7). *GI registration for three farm products*, available at https://www.thehindu.com/todays-paper/tp-national/tp-kerala/GI-registration-for-three-farm-products/article15678026.ece. Accessed 3 Dec 2019.

Thrupp, L. A. (2000). Linking agricultural biodiversity and food security: The valuable role of sustainable agriculture. *International Affairs, 76*(2), 265–281.

Zhu, Y., et al. (2003). Conserving traditional rice varieties through Management for crop diversity. *Bioscience, 53*(2), 158–162.

# Chapter 11
# Community-Based Self-Help Groups in Agriculture

**Beena Anil, Matthew Tonts, and Kadambot H. M. Siddique**

**Abstract** Traditionally, government extension agencies have played a critical role in ensuring that farmers had the knowledge, skills, and competencies to remain competitive and sustainable. However, with the decline in public sector extension activities across much of the developed and developing worlds a greater input is required from the private sector and farmers themselves. This chapter provides background detail on the developments that led to the emergence and evolution of grower groups as crucial players in the agricultural paradigm and describes the critical role played by these groups in sustainable growth and development. It details the development of self-help groups in India.

**Keywords** Extension activities · Collaborative · Participatory group processes · Self-help groups in India

## 11.1 Introduction

Agriculture, in its business-as-usual mode, is by nature very adaptive. Agriculture in the twenty-first century faces multiple challenges as it has to (1) cater to the needs of a growing population with a smaller rural labour force, rising input costs and land degradation, (2) contribute to the overall development of many agriculture-dependent developing countries, and (3) adopt efficient and sustainable production methods to respond to the challenges of climate change. To remain competitive,

B. Anil
The UWA Institute of Agriculture, The University of Western Australia, Crawley, Australia
e-mail: kadambot.siddique@uwa.edu.au

M. Tonts
Faculty of Arts, Business, Law and Education, The University of Western Australia, Crawley, Australia

K. H. M. Siddique (✉)
Hackett Professor of Agriculture Chair and Director of The UWA Institute of Agriculture, University of Western Australia, Crawley, WA, Australia



M. Blakeney, K. H. M. Siddique (eds.), *Local Knowledge, Intellectual Property and Agricultural Innovation*, https://doi.org/10.1007/978-981-15-4611-2_11

farmers need to engage in a process of ongoing adaptation, ensuring efficient use of technologies and practices, while considering long-term environmental, social and economic sustainability.

Traditionally, government extension agencies played a critical role in ensuring that farmers had the knowledge, skills, and competencies to remain competitive and sustainable. However, since the 1980s, the centralised, state-led extension activities across much of the developed and developing worlds have been cutback in favour of approaches that involve more input from the private sector and farmers themselves. Enhanced farmer participation in the generation of knowledge and improved practices is now accepted as highly effective in supporting fundamental changes to farm management practices, thus promoting more sustainable systems. The retreat by central governments from research, development and extension in agriculture, and thus the increasing focus on farmer-led participatory strategies, has contributed to the emergence of farmer-based organisations that have a critical new role in the promotion of more sustainable agricultural systems. While the group-based approach is not a new phenomenon in agriculture, and organisations such as farmer's co-operatives, machinery pools and credit unions have a long history, rarely were they actively engaged in 'grass-roots' level extension, research and development. These organisations tended to serve as lobbying groups and focused on delivering economic benefits to the members, not addressing issues on a system-wide basis. In contrast, many of the newer groups not only focus on production improvements, but also environmental protection and rehabilitation, poverty reduction, and social sustainability. This reflects the growing recognition that the development of more sustainable agricultural systems requires integrated approaches that not only involve high levels of community or farmer participation, but also engage with the complex economic, social and environmental processes.

This chapter provides background detail on the developments that led to the emergence and evolution of grower groups as crucial players in the agricultural paradigm and describes the critical role played by these groups in sustainable growth and development.

## 11.2 Changing Perspectives on Extension

For almost three decades, public-funded extension remained the primary source of information dissemination in agriculture. Publicly-funded extension started by adopting the 'top-down, linear model' of technology dissemination, and experimented with progressive modifications of the model for technology transfer and influencing the adoption behaviour of farmers (Black, 2000; Fulton et al., 2003). Significant investments were directed into public sector agricultural research and extension services to ensure that farmers had sufficient access to inputs and technical information to improve their production and profitability. This led to extension being linked to specific capital investment. For countries like Australia and New Zealand, the 1970s witnessed the zenith in extension services in terms of government resources committed. These rapid investments were mainly seen as a means to improve the export potential of agricultural commodities (Cary, 1993). For developing countries, accelerated public sector investment aimed at improving food production and rural development continued until the late 1980s. The model contributed to a noticeable increase in agricultural productivity and profitability. However, over time, there was a growing realisation of the shortcomings of the very large but often haphazard investments made in both public sector extension and research, and the extent to which they were having an impact.

Meanwhile, rapid globalisation contributed to extensive changes in the agricultural sector (Pritchard & Tonts, 2011). During the 1980s and 1990s, farming became increasingly integrated into global supply chains and is now characterised by increased global competition, industrialisation, precision (information-intensive) production, and risk and diversity. There is also more emphasis on environmental sustainability. Large-scale multinational companies (supermarket chains, meat processing units and community firms) now dominate the worldwide food industry within the agricultural supply chain. Major life science companies (e.g. Syngenta, Monsanto) dominate the input-supply industry, like seeds and agrochemicals. These changes at both ends of the global food supply chain have steadily contributed to a more complex agri-food system (see Argent, 2011). This resulted in greater involvement of the private sector, mainly in the agricultural research, product and service delivery (Hall, 2005; Rivera, 2000). These private companies play an active role, especially in the underlying research continuum. For example, many private firms are now actively involved in patenting new gene research and bioengineering technologies. The Food and Agricultural Organisation (1999) reported that approximately 80% of the research in plant biotechnology during the 1990s was done by the private sector. In many instances, the private sector (especially input suppliers and output buyers) also became increasingly active in instructing farmers in the process and standards desired by the markets. Often enough, these information providers created specific demonstration plots and field trials, similar to the public sector extension techniques but with the primary intention of strengthening vertical linkages in the agri-food supply chain.

This increasing dominance of the private sector was also a reflection of the shift in ideology from the public to private sector authority. It was accompanied by a transition to global capitalism and free-market principles. This progressive privatisation resulted in an increasing emphasis on efficiency and the steady commercialisation of agriculture, where production would be mainly driven by market signals (Rivera & Cary, 1997). These changes tended to suppress the real prices of major agricultural commodities, which had direct impacts on farmers, particularly given the rising input costs. While staple food prices continued to fall, the pay-off shifted to agricultural strategies that enhanced diversification and increased value-added farm production. In countries like Australia, where agriculture is mostly export-oriented, farmers sought economies-of-scale through farm expansion and investment in technologies and practices that could improve productivity.

One of the outcomes of the increasing emphasis on efficiency meant increased competition, which ultimately led to the 'commodification' of agricultural knowledge and information. Agricultural knowledge was no longer considered a public good and free of charge. The commodification of agricultural knowledge was considered a major factor in shaping the future role of agricultural extension (Rivera,

2008). Around the same time, rapid improvements in information and communication technology occurred, with the internet developing as a global network of knowledge transfer and sharing. This rapid growth in information technology meant that the shift towards private hegemony was accompanied by another power shift. It involved the weakening of national boundaries by international electronic technology and trade liberalisation, suggestive of a move away from national limits to global systems of interaction (Rivera, 2008). This meant agricultural knowledge could be transferred more easily than before.

The combination of these economical, policy and technological changes meant that national governments in many countries were under pressure to reform the public sector extension systems. These changes favoured the downsizing of public research and extension services and the emergence of new forms of public/private partnership, the privatisation of parastatals, and increased overseas private investments. The impact of these changes on extension policies was severe. Extension programs across the world increasingly faced the challenge to be relevant and practical and to create a desirable impact in a rapidly changing world. In a report released by the Extension Committee on Organisation and Policy (ECOP) in 2002, the impact of globalisation was listed as one of the six key challenges facing the extension system. These challenges were accompanied by corresponding changes in the notion of agricultural research.

In the 1990s, agricultural research agendas broadened to include targets such as attaining food security, while also focusing on sustainable resource use and the socioeconomic well-being of farmers. These developments further broadened the area of responsibility for extension at a time when attempts were being made to reduce public sector investment. The combined effect of the above factors led to a critical assessment of extension worldwide (Rivera & Qamar, 2003). The argument was that public sector extension alone would never be able to fulfil the broad range of objectives under these changed circumstances. There was growing uncertainty about what role extension was supposed to play in agricultural and rural development. The principal concern was how to deal with the macroeconomic reforms that were changing the fundamental conditions both for public extension services and farmers as producers (Farrington, 1994).

## 11.3 Reform Initiative: Towards a Participatory and Pluralistic Extension

By the mid to late 1980s, public sector extension had reached a juncture where the possibility of regaining the stature it once held was unlikely. With no prospects of respite from the increasing financial constraints and the pressure to slim down and refocus, most governments instigated a move away from the philosophy 'government must provide' to adopting more practical alternative approaches. Different 'fragments' of innovation reforms were bought about to change the public sector extension system. These included structural changes aimed at the privatisation and decentralisation of extension services, changes in the mode of funding involving cost recovery, organisational and management changes, including better linkages with research and greater use of information technology (Rivera & Sulaiman, 2009).

Some changes were also implemented in the extension programs themselves, including better attempts at linking farmers to markets, extension playing a brokering role with the different actors involved in agricultural innovations, and involving stakeholders in development initiatives. The scope of extension was further broadened to include environment and community issues to be more relevant to the needs of rural people. These policy reforms led to structural changes, such as greater participation of local government in the process of financing and managing services, and shifting extension from central to sub-government institutions. The main objective was to improve institutional responsiveness and accountability. Much more emphasis was placed on farmer participation in program planning and implementation, and the involvement of a range of actors and institutional options for financing and delivering extension services. It was also increasingly emphasised that, to gain a better understanding of the extension advisory service, it was crucial to consider it as a component of a broader system of knowledge generation, exchange and use in the agricultural sector.

One of the outcomes of this was the introduction of concepts like the Agricultural Knowledge and Information System (AKIS) and the Agricultural Innovation System (AIS). Both of these focused on better linkages between organisations and the actors involved in facilitating innovation. Under these approaches, farmer participation rapidly gained further legitimacy. They were seen as the perfect means for enhancing the effectiveness of technology generated and diffused, thus promoting greater adoption by farmers. Farmers were also regarded as an excellent platform for improving rural community links with public and private service providers (Sulaiman & Hall, 2002).

## 11.4 Emergence of Agricultural Extension as a Knowledge and Information System

AKIS emerged as an alternative model, promoting better linkages and communication between the system's actors. It followed the notion that knowledge creation, dissemination and users of agricultural knowledge system are bound together and cannot be thought of in terms of 'watertight' compartments. The AKIS combined agricultural research, extension and education in one system[1] and focused on how new knowledge and information could be generated for farmers (Anandajayasekeram, 2011).

---

[1] A set of institutes becomes a system when its individual components are interlinked or articulated, and the separate institutions are connected so that they communicate and cooperate in action to share their human, physical and financial resources in order to achieve one or more common goals.

AKIS primarily focused on technological innovations that were strongly related to the major issues of knowledge and institutional change (addressed within the concept). Although still focused on research supply, much attention was given to identifying farmers' needs for new technologies. It especially emphasised the role of farmers, both as recipients and originators of knowledge and information, in the process of agricultural development. The AKIS concept described a two-way flow of information and knowledge among the research, extension organisations and farmers. Public institutions played an influential role in the innovation process, and group approaches steadily gained popularity as an effective means in the knowledge-generation process. The increasing emphasis on the group approach in countries like Australia is associated with this paradigm (World Bank, 2007).

The knowledge and information system model within AKIS underwent several iterations (Leeuwis, 2010). This resulted in a constant evolution of the perspectives of the AKIS concept over time. One of its initial shortcomings was an excessive focus on large farmers, particularly their technological needs, often at the expense of considering the needs of farmers with limited land and resources. Also, at a time when there was an increasing focus on market forces and the need to engage in global supply chains, AKIS was criticised for its limited attention on the role of commodity markets. It also tended to overlook the heterogeneity evident among farmers, not merely in terms of their agricultural system, but also in terms of social, cultural and psychological diversity. This restricted its scope and limited its ability to offer a complete and realistic framework for research and extension (Leeuwis & van Den Ban, 2003).

In order to extract greater use of knowledge, a much broader set of conceptual and methodological approaches was needed that emphasised the involvement of the entire set of organisations and actors in the innovation generation process, and focused on the role of markets. This became even more crucial with the significant influence of rapidly changing world-economic and social development. Knowledge is a pivotal resource for remaining competitive, and there have been radical developments in the way knowledge is generated.

Unlike traditional approaches, wherein knowledge production was concentrated within scientific institutions (universities, government institutions and industrial research labs) and structured by scientific disciplines, its locations, practices and principles are much more heterogeneous today (Hessels & van Lente, 2008). Their focus on research as the key to generating knowledge shifted to search and consultation. This gave rise to the AIS perspective, which mainly deals with a system made up of innovations occurring on different knowledge fronts, such as the formal research systems, the private sector, farmers and other social actors in the broader policy, cultural and institutional environments.

## 11.5 The Emergence of Agricultural Innovation Systems

Innovation system thinking provides an analytical framework that explores complex relationships among heterogeneous agents, social and economic institutions, and endogenously determined technological and institutional opportunities. Under this model, innovation is seen as neither science, research, nor technology, but rather the application of knowledge to achieve desired social and/or economic outcomes. The innovation system thus extends beyond the creation of knowledge to include factors that affect the demand for and use of new and existing knowledge in novel and useful ways. The concept is robust: its principles are derived from direct observations of countries and sectors with strong track records of innovation, governed by the rules of a free market economy democratic governance system.[2]

Most of the observations come from developed countries and the industrial sector. It relies little on the public sector for knowledge and mainly on the private sector taking the lead with technological innovation. Within the agricultural sector, the application of innovation systems has evolved in two ways: (1) as a framework for organisational analysis, and (2) as a framework for technology development and dissemination. On the organisational side, it began with National Agricultural Research Institutes (NARIs)[3] and led to AISs, including all organisations focusing on knowledge generation, dissemination and application. On the knowledge side, it moved from near technology and productivity gains to broader organisations focusing on knowledge generation, dissemination and application. On the knowledge side, it moved from near technology and productivity gains to broader development goals of poverty alleviation, food and nutrition security, and environmental sustainability.

The AIS concept exhibits similarities with the multiple source model[4] and adopts a holistic approach with an emphasis on a social network that connects research to knowledge users. The learning process adopted is context-specific and, consequently, institutional learning can lead to a great diversity of approaches, partnerships and strategies. Explicit use of the innovation approach is now being made by several policy analysts in relation to agricultural knowledge and technology generation, especially in the developed world. Though the application of this concept is relatively new, it is increasingly suggested as a way of revisiting the question of how to strengthen agricultural innovation capacity.

## 11.6 Participatory Approaches: Capacity Building through Interactive Social Learning Processes

One of the logical extensions of the paradigm shift from the linear model to multiple-source models of innovation like AKIS and AIS has been an emphasis on participatory approaches in the agricultural knowledge generation and dissemination process

[2] Democratic governance includes the separation of powers and independence of the branches of government, the exercise of powers in accordance with the rule of law, respect for human rights and fundamental freedoms, and the transparency and accountability of a responsible civil service, functioning at both the national and local level.

[3] The framework for NARI emerged after World War II that facilitated major investment in agricultural research to improve food production.

[4] In the multiple source model, major emphasis is given to the idea that innovation comes from multiple actors including researchers and practitioners.

(Anandajayasekeram, 2011; Asenso-Okyere, Davis, & Aredo, 2008; Marsh & Pannell, 2000). Over the years, participatory approaches in agriculture have undergone profound changes both in their objectives and implementation. Originally developed as a means to adapt technologies to farmers' local conditions, their focus later shifted to farmer capacity building and empowerment. With the recent emphasis on innovative systems thinking, these underwent yet another adaptation and are now seen as a means for developing more explanatory solutions and innovative results to deal with the complexities involved in modern agriculture (Black, 2000).

A collaborative, participatory group process is adopted in pooling together skills, knowledge, experience and other resources from the multiple partners involved in the development of more context-specific information. The commitment to participation by farmers is more than a form of 'tokenism' and an essential part of the process of learning about needs, opportunities and the actions required to achieve them (see Arnstein, 1969). Most of these approaches allow greater farmer initiative for identifying problems, and planning and implementing research and development activities. In most cases, project interventions are the means to promote collaborative, interactive and experimentation methods of learning. The interactive learning approach followed within the participatory framework, while strengthening farmers' planning and management abilities, builds on their knowledge and practice, thus promoting their capacities to develop and adopt new and appropriate technologies.

A corollary to this increasing emphasis on participatory approaches is the growing significance of the notion of 'social learning', which now is seen as a central theme in the literature on agricultural development (Morgan, 2011; Munshi, 2004; Oreszczyn, Andrew, & Susan, 2010). It is becoming a normative goal in the context of agricultural development projects and policies, although there remains little consensus over its meaning or theoretical basis. Social learning in the present context of knowledge generation is closely associated with members working together in a group to develop understanding, knowledge and skills in collaborative and collective ways, mainly by using and contributing to a range of shared resources (van Buuren & Edelenphos, 2006).

## 11.7 Self-Help Groups in India

In the case of India, where around 70% of the population lives in rural areas, the development of more sustainable agricultural systems is a significant policy concern. As in many other parts of the developing world, farmers are facing a range of challenges associated with cost-price pressures, climate change, knowledge and skill deficits, and difficulties accessing the latest technologies. Until recently, the government played a major role in the economic development of the country, particularly as it relates to agriculture. The main strategy was direct participation by the government in economic activities, such as production, marketing, research and development, and extension. The government also regulated private sector economies through a complex system of controls. In addition, the Indian economy was sheltered from foreign competition through the use of both the "infant industry argument"[5] that provided price support for some agricultural commodities, and a binding foreign exchange constraint. The high level of subsidies provided by the government led to a payment crisis by the early 1990s, and a series of economic reforms. There was a move away from the centrally planned development strategies to those based on market-led approaches. This is consistent with the neoliberal policy reforms experienced in many other parts of the world during the 1980s and 1990s that involved the deregulation of economies, privatisation of state enterprises, and focus on the private sector delivering many vital services. One of the other key reforms was an emphasis on the promotion of entrepreneurial activities through community participation. Policymakers in India now regard community participation as a vital strategy to work with rural people in improving their economic and social status. In the case of Indian agriculture, community participation is regarded as a critical ingredient in the pursuit of farming systems, which are not only more economically viable, but also socially and environmentally sustainable. Accordingly, the promotion of community- and farmer-based self-help groups in agriculture is one strategy that has been adopted by the government. These groups can play a significant role in many core aspects of farming, such as increasing production at a reduced cost, providing expert technical guidance, purchasing inputs, marketing products, training, credit or equipment, representing members' interests, building influence, fundraising, and carrying different projects. The activities of two self-help groups, Confederation of Potato Seed Farmers (POSCON), operating at a state level, and the Global Self-Help Group, a local level group are described to further explain the role of the farmer groups in the current agricultural scenario in India.

## 11.8 Confederation of Potato Seed Farmers (POSCON)

POSCON was initiated in 2007 in the northern state of India, Punjab, following the intensive effort of a few growers to bring all potato seed producers within the state under one umbrella. In 2008, the group was registered as a society within the state; it is the only potato crop organisation affiliated with the Government of Punjab. The group's primary objective is to promote, develop, build and propagate seed potato cultivation for the benefit of seed potato growers in Punjab by adopting and applying the most advanced and modern technologies. It aims to produce and market the best quality potato seed to compete with the best seed potato growers in the world. Several initiatives were put forward by the group for cultivating potato, marketing, developing linkages with relevant agencies such as agribusiness dealers, and promoting exchange programs for members. The group has links with financial

[5] The infant industry argument refers to a development approach that favours supporting new or "modernising" industries through various forms of production, subsidy and intervention.

institutions, farm equipment manufacturers and other agencies supplying various farm inputs. Every effort is made to arrange farm inputs at below-market price. Members of the association sell their produce in an individual capacity.

POSCON started with 163 members in 2008, with membership steadily increasing to >250 in 2010–2011. All members constituting the General House meets once a month and discuss common issues relating to members' interests. The monthly meeting has approximately 70% attendance. The group leader is selected by consensus for a term of 2 years. The group employs an office secretary to take care of office work. The executive committee of the group comprises 14 members who meet once a month and when required. The executive members are responsible for marketing the group's produce and managing savings. The yearly membership of the group costs 5000 Indian Rupee (INR) for ordinary members and 10,000 INR for executive members.

Members of the group meet regularly; group activities, such as field trials and general group meetings, form the basis of group learning. Seminars and conferences are also organised, where presentations by experts (scientists/government officials) play a key role in the exchange of information between farmers and experts—the farmers learn the latest in research and the experts learn more about local farming systems and the constraints and challenges encountered by growers in their day-to-day operations. This helps farmers and researchers to use resources more effectively and build capacity. Representatives from the group regularly participate in Global Potato Conferences and the World Potato Congress, which serve their need to exchange ideas and information at a global level. Also, the group has signed a Memorandum of Understanding (MoU) with Farm Technology Network (FTN),[6] the USA on technology sharing, particularly in the area of soil nutrients and water management.

POSCON members produce and market between 50% and 60% of the potato seed requirements of India. The produce is marketed both within and outside the state and the varieties preferred by the group are those approved by the Government of India. The group has established contacts with a wide range of stakeholders, and regular meetings are held between group members and the Chief Minister of the state and officials from the Department of Horticulture, Department of Agriculture, Electricity Board, Punjab Marketing Board and Punjab Agricultural University. The group has also established partnerships with several private organisations, including tractor companies, potato machinery and equipment companies, and plant protection companies. These partnerships focus on finding solutions to problems faced by group members in developing productions systems appropriate to local conditions, through expert guidance, skill-oriented training, learning from each other and participatory research. Occasional advice is sought from agricultural institutions. Members of the group are regular subscribers to the global gap certification process, which makes them eligible for potato export. One of the strengths of the group is its

[6]FTN is experienced in working with growers on soil fertility issues with more than 40 different crops and works within the USA and 15 countries worldwide.

ability to lobby various input-supply companies and semi-government departments and corporations to safeguard the interests of its members.

One of the successes of the group is the high level of member participation in group activities to achieve group goals (Kalra, Anil, & Siddique, 2013). Increased income, availability of advanced technical information, social networking and transparency in functioning were identified as the facilitating factors impacting member participation. At present, this group is in a position to help other groups, especially those with members with lower socioeconomic status. Group activities, opportunities and challenges are summarised in Table 11.1.

## 11.9 Global Self-Help Group

The group was started voluntarily by 16 women in 2008 to socialise and achieve economic self-sufficiency. A president, secretary, cashier and two council members were chosen to form an executive committee. Group meetings occur twice a month and are attended by almost all members. Apart from discussions on production and preparation of value-added products, group members socialise through activities, such as folk dance, folk song, poetry recitation, and planning for festival celebrations. The group is financed by small monetary contributions from members and a loan from a cooperative society. Members unable to make a monetary contribution to the group provide in-kind efforts by working extra hours for the group. Group members are involved in the production of organic seasonal vegetables, honey

**Table 11.1** Summary of POSCON group activities, linkages, impact, opportunities and challenges

| | Group type | Linkages | Group management | Impact | Opportunities and challenges |
|---|---|---|---|---|---|
| POSCON | Community-based learning group | Government organisations, research organisations, financial institutions, agribusiness, other farmer networks at national and international level | General House—comprising all the members—determine the direction of the group.<br>Executive Committee of growers manages finances and group activities.<br>Group administration undertaken by staff. | Influence member learning and adoption, mainly through participation in group research and development activities | **Opportunities:** Active member participation in research and development activities comparatively lower.<br>Ability of the group to lobby various input-supply companies, semi-government departments and corporations to safeguard the interest of its members. |

value-added products[7] and north Indian delicacies. Raw materials are procured from members' farms and other farmers in the village.

The group has established linkages with Punjab Agricultural University (PAU), Central Institute of Post-Harvest and Engineering Technology (CIPHET) and other development departments. To market its products, the group has linkages with other organisations, including the Department of Agriculture, Horticulture and Animal Husbandry and Farm Science Centre/Krishi Vigyan Kendras (KVKs) of PAU. The group manages its marketing, with savings usually invested back into the group's business or used for inter-loaning among group members at 12% interest with 1% interest returned to the pool. The group regularly takes part in national and state events; for instance, the group has attended farmers' fairs and state-level workshops and seminars to upgrade their knowledge and skills. Thus, the group plays a critical role in empowering women by creating new economic opportunities and enhancing social support networks.

One of the unique features of the group is that even though only a few members play an active role in the development and promotion of group activities, all members take part in these activities. Group members share extremely close social, cultural and family ties. Members of the group are well-coordinated and manage the group activities without any paid staff. Since its inception, the group has steadily improved its profit margins and won many state-level awards and small-scale projects at the national level. The timely provisions of grants and aids helped the group to build adequate infrastructure to streamline its activities and aid in its establishment. The collective efforts of group members, active participation in group events, shared interests and transparency in accounts are some strong points of the group.

At present, this group is in a position to help other existing groups, especially those with members of lower socioeconomic status within the village and neighbouring villages. Group activities, opportunities and challenges are summarised in Table 11.2.

**Table 11.2** Summary of global self-help group activities, linkages, impact, opportunities and challenges

| | Group type | Linkages | Group management | Impact | Opportunities and challenges |
|---|---|---|---|---|---|
| Global self-help group | Community-based group to achieve economic self-sufficiency and meet social needs | Government organisations, research organisations, financial institutions and private organisations | Group leaders provide direction to the group. | Influence member capabilities to achieve economic self-sufficiency by organising training and finances | **Opportunities:** Active member participation in all group activities |
| | | | Group activities managed by group members. | | **Challenges:** Many members work overtime |

[7] Pickles, jams, squashes, ground spices

### 11.9.1 Self-Help Groups in Australia

The past century or more has been one of constant change in the funding and delivery of extension services to farmers in Australia. This is particularly true for the past four decades, which have witnessed several policy innovations related to agriculture and more specifically to extension. A significant consequence of these policy changes in extension was a move away from it being largely a public sector activity, focusing on production-based, one-on-one technology transfer, to a group-based approach with multi-stakeholder involvement in the promotion of 'market-based' agricultural information.

Extension in Australia, from the very beginning, was influenced by the environment in which farmers and extension agencies operated. Much of the momentum for early government extension was problem-centred and focused on overcoming challenges associated with the alienation of undeveloped land. After overcoming initial settlement challenges by the 1940s, there was a subtle shift in government policies that focused less on land alienation and more on increasing agricultural productivity. Productivity-based technology transfer soon assumed the predominant focus of agricultural extension. The outcome was a rapid improvement in agricultural productivity and profitability (Williams, 1968). By the end of the 1950s, agriculture was contributing close to 85% of export earnings and nearly 20% of the GDP.

An important dimension of agricultural development policy around this period was the application of economic principles of farm management. Farm-level problems were often tackled using insights from economists as a means of identifying principles on which efficient farm management depends. Soon these economic theories and concepts were accepted as part of the nature of advisory services being provided to farmers, under the rubric of farm management extension. Several initiatives were taken up by state departments and agricultural faculties of universities to incorporate farm business management information in their routine agendas. Many programs were designed to educate groups of farmers about the management of their farms. However, comprehensive supervisory farm management services for individual farms limited the number of farms that benefited from public sector extension programs. This provided the needed incentive for the private sector, especially agricultural management consultants, to deliver advisory services to farmers (Gray & Lawrence, 2001).

By the 1960s, private consultants were seen as significant source of information and dealt with the whole farm on a management as well as technical basis (Schapper, 1962). Another significant development was the increasing influence of the agribusiness sector in agriculture and related policies. The period of rapid agricultural progress was followed by accelerated adoption of technologies and specialisation in agriculture, which resulted in the increased involvement of corporate agribusiness firms, particularly the supply of chemicals, seeds and fertilisers. Some of these commercial service providers took an active interest in providing advice to their farmer clients. Their role in extension was essentially confined to commercial purposes and was mainly meant for promoting their products and services among their clients.

Another development during this time was the increased financial contribution by the Commonwealth government to improve extension services for almost all rural industries. A further addition was increased farmer participation in the development of technical services for rural industries through their involvement in primary industry organisations. All these developments led to rapid growth in agriculture and most farmers experienced unprecedented prosperity. The rapid mechanisation of farming during this era, while contributing to an accelerated rise in the cost of farming, also led to steady growth in farm holdings and improved efficiency (MacDonald, 2011). With farms getting larger and highly specialised, the need for more complex, individually tailored, technical, management and marketing information increased further which reinforced the role and contribution of the private sector in the business of agriculture.

Agriculture as an industry continued to grow until 1980; after that, the rapid growth slowed and there was a steady decline in agriculture's contribution to GDP. The 1980s also witnessed a steady decline in the price of major agricultural commodities due to global overproduction, and the price of inputs rose steadily. This led to a severe 'cost-price' squeeze, and many farmers with small enterprises struggled to remain viable. The combined impact of these, together with diminishing levels of political support, resulted in lower levels of public investment in agricultural development and extension. Strategies developed during this time, essentially encouraged the exit of small, inefficient producers from agriculture, while encouraging larger producers to expand their operations and take advantage of scale economies.

The emergence of neoliberal policies during the 1980s saw an increased focus on microeconomic reforms, which emphasised a shift from state paternalistic policies to a more market-oriented position. As part of an increased emphasis on agricultural deregulation, public sector services like extension were under severe scrutiny. An important question being asked was: could extension not be left with the private providers? Why was government involvement needed? Australian agriculture during this era moved from a complex array of government interventions to one of the most economically liberal in the world. As part of the broader policy framework to improve efficiency, governments also increasingly promoted the concept of 'self-reliance' in agricultural and regional policies. Rural producers were told to become less dependent on government support and develop a more entrepreneurial attitude to financial and environmental risk.

In many respects, Australia had a head start over many other countries in thinking about and implementing change in the public funding and delivery of extension. The government adopted varied innovation reforms in their public sector extension systems that resulted in a reduction in direct government provision of extension services. There was also a steady increase in private sector involvement in the funding and delivery of both agricultural research and extension. These developments meant a greater reliance on participatory approaches as the preferred means of information generation and dissemination and, consequently, a greater emphasis on interaction and partnership between farmers and the broader agricultural industry.

The departments of agriculture moved away from their traditional production-based information dissemination role and adopted strategies to provide a more accountable, participatory and business approach focused on the market and customers, which meant they had explicitly stated objectives of acknowledging their client needs. Some set up formal links with industry in an attempt to ensure that industry needs were met. Various interpretations of the Funder–Purchaser–Provider (FPP) model,[8] as an organisational model, are being implemented by the departments of agriculture in most states. This model allows for 'activity-based' or accrual auditing, seen by its advocates as being important to improve accountability. As the principles of the model are implemented and outsourcing of activities increasingly occurs within public sector agencies, opportunities for private consultants and agribusiness to deliver government-funded research and extension programs have improved. Regionalisation[9] is also being promoted to varying degrees in each state to promote integration across agency programs and negotiation with clients on better service provision. Agricultural departments are also adopting 'user-pay' philosophies, charging for information delivery. Under this approach, Research Development Centres (RDCs)—representing major sectors of commodity production—emerged as a major player in agricultural research and extension activities.

Some of the outcomes of the increasing influence of the RDC model were: (1) greater allocation of funds for demand-driven research and extension that matches the RDC's priorities and an increased accountability need; (2) demand for more cross-industry and national agendas of research and extension initiatives forcing the need for better public/private sector liaisons; (3) increased questioning about infrastructure costs, and 'in-kind' contributions estimated by government organisations, including the universities and CSIRO; and (4) demand for more private sector participation in research and extension programs. These changes required the involvement of a complex web of providers and investors with reliable interconnections in agricultural, research, development and extension activities within Australia. Consequently, participatory group-based approaches gained prominence as the accepted model for agricultural research and extension activities. These approaches while favouring multi-stakeholder partnerships, emphasised increased integration and learning between different disciplines and greater involvement of farmers in the generation and dissemination of relevant information.

Farmer groups were increasingly accepted as a means for the production and dissemination of farmer-relevant information. From the 1990s, various brand name groups (Target 107, Topcrop8) were promoted by RDCs primarily to disseminate the results of their funded research and mainly seen as sources of information retailing. Local 'best practice' groups were also increasingly promoted in some states

[8] Funder–Purchaser–Provider model requires the separation of policy and service delivery responsibilities. The implementation of this model involves effective benchmarking of the cost of services.

[9] Regionalisation is a policy where regional managers act as providers under the FPP model to ensure integration across the State Department of Agriculture's programs and negotiate with clients on service provision.

like Queensland. These groups emphasised participatory learning as a means to gaining an understanding of relevant problem issues. Various group approaches, focusing on human resource development, were also promoted. For instance, the Farm Management 500 (FM500) program commenced with the broad objective of helping farmers to improve farm management practices. The main emphasis of this program was learning through interaction with others in the group (Anil, Tonts, & Siddique, 2015a).

The early 1990s also witnessed the emergence of farmer initiated and managed production-focused groups throughout Australia. These are community-based groups of farmers who apply local knowledge, together with support from government agencies, to focus on production issues at local and regional levels (Anil, Tonts, & Siddique, 2015b; John, 2005; Lave & Wenger, 1991). The farmer group movement in Australia has, in many instances, gone beyond the range of international participatory research and extension models and has redefined 'participation'. For some of these groups that employ their researchers and directly receive significant research and development funding, the concept of participation becomes very different. In some cases, it is the traditional research organisations and researchers that are now encouraged to become participants at the invitation of the farmers. The research focus of the group is usually defined by the objectives of the group, and generally designed by the members with assistance from a facilitator or scientists. Together the farmers and scientists are involved in the research—the results of which are communicated to farmer members and the broader industry through various group events and publications, such as field days, crop updates and newsletters. The groups contribute to research and extension by identifying field sites for field trials, collaborating as partners in large projects and, in some cases, developing and implementing new projects with research providers. Increasingly seen as an important means for information dissemination, RDCs are making huge investments in grower group-based trials and further extension of results.

## 11.10 Western Australia No-Till Farmers Association (WANTFA)

WANTFA is the largest agronomic grower-initiated group in Western Australia, formed in the early 1990s to address widespread erosion problems within the state. The group was initiated by a small network of people who believed that no-till technology could help address erosion problems effectively. Within just 3 years of its formation, the group developed its first no-till seeding equipment in partnership with a local manufacturing company, and this marked the start of a rapid growth phase in the life of WANTFA. By the fifth year, group membership grew to >600 people, and in the same year, a significant breakthrough occurred with the establishment of a long-term partnership with Grains Research and Development Corporation

(GRDC).[10] With funds from GRDC, the group appointed its first staff and also established a technology demonstration trial site in the central grain belt region of the state to conduct long-term research to assess the benefits of the no-till system. Once established, the WANTFA trial site became the centrepiece for most of the group's subsequent projects and trials. Together with investment from GRDC and in partnership with other government and private sector organisations, WANTFA carried out high profile no-till demonstrations and extension activities. The group membership rapidly rose to around 1400 members, spread across the grain belt region of the state along with some interstate and overseas members.

The group was managed by a team of growers with administrative and technical support provided by core staff. The funds for the group were mainly received from GRDC, Australian Government Department of Agriculture, Fisheries and Forestry, National Landcare Program, corporate sponsors, events income and membership fees. The rapid rise in WANTFA membership was accompanied by a steady rise in no-till adopters within Western Australia. Between 1998 and 2009, the adoption rate of no-till farmers within Western Australia rose from 25% to 90%. This accelerated rate of adoption was rapid by extension standards (typical changes in agriculture takes 10+ years), and WANTFA is considered to have played a significant role.

With most farmers within the state having adopted no-till practices, the group extended its focus to conventional farming techniques and refined its management structure. The group is managed by a board comprising both farmers and scientists with administrative and technical support from the staff. The group has membership with other grower networks such as Grower Group Allianz (GGA)[11] and the Conservation Agricultural Alliance of Australia and New Zealand and has established partnerships with various government and private organisations. Learning within the group occurs mainly through participation in group activities, such as field trials, field days, seminars, conferences and study tours. The group magazine also serves as a crucial source of information among group members and the wider community. WANTFA, through their broader reputation as experts in their field and their well-established networks, are in a position to play a key role in carrying knowledge on conservation agriculture beyond the boundaries of its membership at a national and international level. Group activities, opportunities and challenges are summarised in Table 11.3.

## 11.11 Mingenew Irwin Group (MIG)

MIG is a regional group based in the northern agricultural grain belt of Western Australia, covering approximately 300,000 hectares across the Mingenew and Irwin areas. It was officially formed in 1998 by merging two Land Conservation District

[10] The GRDC is responsible for planning, investing and overseeing research and development, and delivering improvements in production, sustainability and research and development across the Australian grains industry.

[11] The GGA is an umbrella organisation connecting grower groups within Western Australia.

**Table 11.3** Summary of WANTFA group activities, linkages, impact, opportunities and challenges

| | Group type | Linkages | Group management | Impact | Opportunities and challenges |
|---|---|---|---|---|---|
| WANTFA | Community-based learning group | Government organisations, research organisations, funding bodies, NRM bodies, agribusiness, other grower groups within the state and interstate groups, international farmer networks | Board comprising growers and scientists. | Influence member learning and adoption mainly through participation in group research and development activities. | **Opportunities:** Have wide coverage |
| | | | Group administration undertaken by staff. | | **Challenges:** Active member participation in research and development activities comparatively low. |
| | | | | The group magazine is a key source of information for many interstate and overseas members. | Wide membership base limits addressing broader objectives. |

Committees and focused on integrating farm production and 'landcare' to encourage economically and environmentally sustainable agriculture in the region. From there, the group grew slowly reaching a membership close to 100 by the tenth year and grew rapidly to nearly 200 members in the next 4 years. As part of its research and development activities, the group organises nearly 70 small plot trials, farmer demonstrations, and independent trials for product suppliers and on ground natural resource management work. The MIG is considered one of the most successful farmer groups within the state and was awarded the 'Innovation in Sustainable Agriculture' award as part of the National Landcare Awards in 2010.

From the beginning, MIG has focused on being predominantly grower-led and managed, while being strictly confined to its region of operation and accommodating newer aspects as part of negotiated goals. This has played a key role in influencing the group structure and mode of operation and considerably influences the mutual commitment of members to group research and development activities and the strong bonds between them. A grower-led management committee determines the direction of the group. The committee meets once a month, except during the peak seasons of seeding and harvest. Members often nominate themselves for positions within the management committee; however, the final decision is by an election at the annual general meeting of the group.

Operational support for the group is provided by its Research and Development Division, while the Environmental Division is driven by farmers. The division committee meets regularly with project staff to provide feedback on progress, develop new themes and ideas for upcoming projects, and ensure that ownership of growth and direction of the group is managed by the members. The projects, trial works and extension activities are then jointly undertaken by group members and staff. Thus, the joint enterprise—centred on the process of creating and disseminating information for agricultural development in their region—builds a mutual interaction among MIG members. This facilitates the management of the group as a social learning structure and plays a key role in influencing member engagement in group activities. The organisational structure of the group also provides a relatively undemanding environment, in communal terms for the group leaders. Leader 'burnout' is a common concern often noted among farming system groups; however, in the case of MIG, members seem more willing to take up leadership roles and commit both time and resources to group research and development activities.

For MIG members, field days are the preferred means of learning. Most MIG field days are well attended. The MIG has established linkages with government departments, research organisations, NRM bodies, funding bodies, agribusiness (especially those in the local region) and the Grower Group Alliance. For undertaking research, the group has a long-term trial site and several satellite sites across its area of operation. Each year, MIG organises nearly 70 small plot trials, farmer demonstrations and independent trials for product suppliers and on ground natural resource management work, with crop updates and a spring field day being the major event. The results of the group's projects and trials are disseminated through group events and publications. The group also has a website providing details about its current information. The key issue each year is identified by conducting a survey and brainstorming session among members. Once identified, the research and development committee is responsible for prioritising the issues and developing trials or collaborative demonstrations to address key concerns. Most of the funds for group activities are obtained through sponsorship, membership fees and investments by project partners. The group events are regarded highly by group members. The MIG is considered one of the most successful grower groups in Western Australia and is considered to play a crucial role in the generation and dissemination of farmer-relevant information. Group activities, opportunities and challenges are summarised in Table 11.4.

## 11.12 Reflection on Community-Based Self-Help Group Activities: Challenges and Opportunities

A community development approach emphasises self-help, the democratic process and local leadership in community revitalisation. Self-help groups are part of the community participation concept which stresses the active participation of members for their welfare. Often, self-help groups emerge from the basic need to address members' economic, social, learning and development needs and present a platform for systematic inquiry, participation and action to address challenges. These groups often focus on sharing best practices and creating new knowledge and act as a forum

**Table 11.4** Summary of MIG groups activities, linkages, impact, opportunities and challenges

| | Group type | Linkages | Group management | Impact | Opportunities and challenges |
|---|---|---|---|---|---|
| MIG | Community-based learning group | Government organisations, research organisations, funding bodies, NRM bodies, agribusiness, other grower groups within the state and interstate groups | Management committee of growers determine the direction of the group. Group administration undertaken by staff. | Influence member learning and adoption mainly through participation in group research and development activities | **Opportunity:** High-level member participation in group research and development activities. With membership confined to the local region, there is much homogeneity among them in terms of their cropping pattern and needs, allowing MIG to focus on a broader range of highly relevant objectives. |

to enable members and stakeholders to come together to explore new possibilities, solve problems and create new, mutually beneficial opportunities. Group membership enables synergy among members for common priority achievement where mutual understanding between members plays a crucial role. Shared values and visions and incorporating group responsibilities, including group leadership, are important for group development and effectiveness. Active participation as a shared vision of action ensures cooperative involvement of members.

The group effort considers the principle that working together enables people to understand each other. It is here that group leaders can play a significant role in a group's success. For group success, the most crucial factor is ownership of the group by its members—right from the start—and constructing appropriate networks. By being part of the group, individuals can access services (e.g. training), credit and marketing linkages. Marketing through groups has clear economic benefits, through the bulk purchase of inputs and access to more information and distinct markets. It also improves the bargaining capacity which helps share risk and cost.

In the case of self-help learning groups, members are brought together by a learning need they share (whether this shared learning is explicit or not, or whether learning is motivated by their coming together or a by-product of it), and undergo collective learning which bonds members over time. Together they are involved in a practice of gathering, evaluating, structuring and disseminating relevant knowledge, which inherently contributes to their learning. Member learning within a grower group is thus substantially influenced by the extent of member participation and interaction. As part of the knowledge gathering process, these groups establish a partnership with the actors and organisations from the broader agricultural industry and collectively undertake numerous projects and trials. The information gathered is disseminated through organised group events and group publications. The group events also serve as an important venue for member interaction which forms a crucial aspect of learning through the exchange of experience, ideas and problems among peers and experts. For most grower group members, participation in group events and interactions are a crucial form of learning. Thus, the ability of groups to provide ample opportunities for member participation and interactions, to a great extent, plays a key role in influencing the effectiveness and sustainability of the groups.

Community-based groups are dynamic and undergo a continuous evolution process. Most groups follow a similar pattern of evolution, typically starting as loose networks that coalesce into a group as members build connections. Often the group leaders play a crucial role in building the initial networks and nurturing the groups. Once formed, a core group of people take up the leadership role in stewarding the group and identifying and defining key domain issues. They begin to create a structure and process for how the group will operate and how the members will work together over time. Sustained interactions among members form the key for developing the trust needed for member involvement and the sharing of information and, over time, the group forms a practice—such as information generation, documentation, organising regular events, establishing norms for management—which sets the rhythm of the group. The steady process of members leaving and new members joining the group, rias and fall in group membership, incorporation of a new domain and change in the structure and mode of operation all form part of the natural process of a group's transformation. In most cases, groups with a well-defined domain, strong member commitment and sustained interactions can survive most transformations.

While most groups appear more or less similar during their initial phase of development, they change in characteristics as they grow. Some set boundaries of operation (local/regional groups), others tend to grow with a broad membership base (state, national and international level). The size of the group can influence member participation and the way groups structure themselves. When size is compounded with distance, it becomes even more significant. For most groups, in their early phase of development, there is a high level of member involvement, especially in group activities (planning and implementing group research, meetings) and events (field days, workshops, conferences, field tours) that slowly diminishes as the group increases in size and coverage.

Naturally, the smaller local and regional groups, with most of their activities (research) and events (field days, workshops, seminars) organised at the local level, invite greater member participation in comparison to the larger, geographically dispersed groups (state-wide). In the case of local and regional groups with members confined to the local region, there is much homogeneity among them in terms of their cropping pattern and needs, which makes it possible for these groups to focus on a broader range of issues. However, regional groups with adequate staff and financial backing have a greater capacity to undertake more research, development and extension activities. For most local and regional group members, group events and interactions form an effective means of learning.

In the case of larger groups, the widely dispersed membership base often results in lower member participation and the variation among members (coming from different regions), in terms of their farming enterprise, makes it challenging to have broader coverage. Often these groups tend to limit their focus on issues not affected by the regional variation. However, larger groups, with their widely distributed membership, tend to make an impact on a wider scale (national and international level).

Notwithstanding these differences, the effectiveness and sustainability of learning-based community groups, to a large extent, depends on the ability of the groups to attract greater member participation in group activities and identify the means to promote regular interactions among group members.

## References

Anandajayasekeram, P. (2011). *The role of agricultural R&D within the agricultural innovation systems*. Paper presented at the ASTI/IFPRI -FARA conference, 5–7 December, Accra, Ghana.

Anil, B., Tonts, M., & Siddique, K. (2015a). Grower group and transformation of agricultural research and extension in Australia. *Agroecology and Sustainable Food Systems, 39*, 1104–1123.

Anil, B., Tonts, M., & Siddique, K. (2015b). Strengthening the performance of farming system groups: Perspectives from a communities of practice framework application. *International Journal of Sustainable Development & World Ecology, 22*, 219–230.

Argent, N. (2011). Australian agriculture in the global economic mosaic. In M. Tonts & M. A. B. Siddique (Eds.), *Globalisation, agriculture and development* (pp. 7–28). Cheltenanham, UK: Edward Elgar.

Arnstein, S. R. (1969). A ladder of citizen participation. *Journal of the American Institute of Planners, 35*(4), 216–224.

Asenso-Okyere, K, Davis, K, & Aredo, D (2008). Advancing agriculture in developing countries through knowledge and innovation, synopsis of an international conference on advancing agriculture in developing countries through knowledge and innovation, 7–9 April 2008, Addis Ababa, Ethiopia.

Black, A. W. (2000). Extension theory and practice: A review. *Australian Journal of Experimental Agriculture, 40*(4), 493–502.

Cary, J. (1993). Changing foundations for government support of agricultural extension in economically developed countries. *Sociologia Ruralis, 23*(3–4), 336–334.

Extension Committee on Organisation and Policy. (2002). *The extension system*. A vision for the 21st century, National Association of State Universities and Land-Grant Colleges, Washington, DC.

Farrington, J. (1994). *Public sector agricultural extension: Is there life after structural adjustment?* (Natural Resource Perspectives No 2) (pp. 4–7). London: Overseas Development Institute (ODI).

Fulton, A., Fulton, D., Tabart, T., Ball, P., Champion, S., Weatherley, J., et al. (2003). *Agricultural extension, learning and change, Rural Industries Research and Development Corporation*. Barton: Rural Industries Research and Development Corporation.

Gray, I., & Lawrence, G. (2001). *A future for regional Australia: Escaping global misfortune*. New York: Cambridge University Press.

Hall, A. (2005). Capacity development for agricultural biotechnology in developing countries: An innovation systems view of what it is and how to develop it. *Journal of International Development, 17*(5), 611–630.

Hessels, L., & van Lente, H. (2008). Re-thinking new knowledge production: A literature review and a research agenda. *Research Policy, 37*(4), 740–760.

John, V. (2005). Community development research: Merging (or submerging) communities of practice? A response to van Valaenderen. *Community Development Journal, 41*(1), 50–64.

Kalra, R. K., Anil, B., & Siddique, K. H. M. (2013). Self-help groups in Indian agriculture: A case study of farmer groups in Punkab, Northern India. *Agroecology and Sustainable Food Sysems, 37*, 509–530.

Lave, J., & Wenger, E. (1991). *Situated learning: Legitimate peripheral participation*. Cambridge, UK: Cambridge University Press.

Leeuwis, C. (2010). Changing views of agricultural innovation: implications for communicative intervention and science. In P. Florencia, G. Singleton, & M. Casimero (Eds.), *Research to impact: Case studies for natural resource management for irrigated rice in India*. Metro Manila, Philippines: International Rice Research Institute (IRRI).

Leeuwis, C., & van Den Ban, A. W. (2003). *Communication for rural innovation – Rethinking agricultural extension*. Oxford, UK: Blackwell Publishing Ltd..

MacDonald, J. M. (2011). *Why farms are getting larger? The case of the US*. Paper prepared for the annual meeting of the German Association of Agricultural Economists (GeWiSoLa), Halle, Germany.

Marsh, S. P., & Pannell, D. (2000). Agricultural extension policy in Australia: The good, the bad and the misguided. *The Australian Journal of Agricultural and Resource Economics, 44*(4), 605–627.

Morgan, S. (2011). Social learning among organic farmers and the application of the communities of practice framework. *The Journal of Agricultural Education and Extension, 17*(1), 99–112.

Munshi, K. (2004). Social learning in a heterogeneous population: Technology diffusion in the Indian Green Revolution. *Journal of Economic Development, 73*, 184–213.

Oreszczyn, S., Andrew, L., & Susan, C. (2010). The role of networks of practice and webs of influencers on farmers' engagement with and learning about agricultural innovations. *Rural Studies, 26*, 404–417.

Pritchard, B., & Tonts, M. (2011). Market efficiency, agriculture and prosperity in rural Australia. In M. Tonts & M. A. B. Siddique (Eds.), *Globalisation, agriculture and development* (pp. 29–53). Cheltenham, UK: Edward Elgar.

Rivera, W. M. (2000). The changing nature of agricultural information and the conflictive global developments shaping extension. *Journal of Agricultural Education and Extension, 7*(1), 31–42.

Rivera, W. M. (2008). The 'business' of the public sector: Extension in transition and the balance of powers. *Journal of International Agricultural and Extension Education, 15*(2), 19–31.

Rivera, W. M., & Cary, J. M. (1997). Privatizing agricultural extension. In B. E. Swanson, R. P. Bentz, & A. J. Sofranko (Eds.), *Improving agricultural extension: A reference manual* (pp. 203–211). Rome, Italy: FAO.

Rivera, W. M., & Qamar, M. K. (2003). *Agricultural extension, rural development and the food security challenge*. Rome: Food and Agriculture Organisation.

Rivera, W. M., & Sulaiman, R. V. (2009). Extension: Object of reform, engine for innovation. *Outlook on Agriculture, 38*(3), 267–273.

Schapper, H. P. (1962). *Farm management club*. Paper presented at the Australian Agricultural Extension conference, Melbourne, Australia.

Sulaiman, V. R., & Hall, A. (2002). *Beyond technology dissemination – Can Indian agricultural extension re-invent itself?* (Policy Brief 16). Delhi, India: National Centre for Agricultural Economics and Policy Research.

van Buuren, M. W., & Edelenphos, J. (2006). Innovations in the Dutch polder: Communities of practice and the challenge of co-evolution. *Emergence, 8*(1), 42–49.

Williams, D. B. (1968). *Agricultural extension*. Melbourne, Australia: Melbourne University Press.

World Bank. (2007). *Enhancing agricultural innovation: How to go beyond the strengthening of research systems, The International Bank for Reconstruction and Development*. Washington, DC: The World Bank.

The manufacturer's authorised representative in the EU is Springer
Nature Customer Service Centre GmbH, Europaplatz 3, 69115 Heidelberg,
Germany. If you have any concerns regarding our products, please
contact ProductSafety@springernature.com

Printed and bound by CPI Group (UK) Ltd, Croydon, CR0 4YY
30/09/2026
02225253-0003